# VALENCE BOND METHODS
## Theory and applications

T0254470

Valence bond theory is one of two commonly used methods in molecular quantum mechanics, the other is molecular orbital theory. This book focuses on the first of these methods, *ab initio* valence bond theory.

The book is split into two parts. Part I gives simple examples of two-electron calculations and the necessary theory to extend these to larger systems. Part II gives a series of case studies of related molecule sets designed to show the nature of the valence bond description of molecular structure. It also highlights the stability of this description to varying basis sets. There are references to the CRUNCH computer program for molecular structure calculations, which is currently available in the public domain. Throughout the book there are suggestions for further study using CRUNCH to supplement discussions and questions raised in the text.

The book will be of primary interest to researchers and students working on molecular electronic theory and computation in chemistry and chemical physics.

GORDON A. GALLUP was born (9 March 1927) and raised in St Louis, Missouri and attended the public schools there. After High School and a short stint in the US Navy, he attended Washington University (St Louis) and graduated with an AB in 1950. He received the PhD degree from the University of Kansas in 1953 and spent two years at Purdue University carrying out post-doctoral research. In 1955 he was appointed to the faculty of chemistry at the University of Nebraska and rose through the ranks, becoming full professor in 1964. He spent a year at the Quantum Theory Project at the University of Florida, and a year in England at the University of Bristol on an SERC fellowship. In 1993 he retired from teaching and since then has spent time as a research professor with the Department of Physics and Astronomy at the University of Nebraska. His research interests over the years include infrared spectroscopy and molecule vibrations, theory of molecular electronic structure, valence bond theory, electron scattering from atoms and molecules, and dissociative electron attachment. During his career he has held grants from the National Science Foundation, the Department of Energy, and others. He has had over 100 articles published in 10–15 different chemistry and physics journals, as well as articles in edited compendia and review books.

# VALENCE BOND METHODS

## Theory and applications

GORDON A. GALLUP
*University of Nebraska*

**CAMBRIDGE**
UNIVERSITY PRESS

CAMBRIDGE UNIVERSITY PRESS
Cambridge, New York, Melbourne, Madrid, Cape Town, Singapore, São Paulo

Cambridge University Press
The Edinburgh Building, Cambridge CB2 2RU, UK

Published in the United States of America by Cambridge University Press, New York

www.cambridge.org
Information on this title: www.cambridge.org/9780521803922

First published 2002
This digitally printed first paperback version 2005

*A catalogue record for this publication is available from the British Library*

*Library of Congress Cataloguing in Publication data*
Gallup, G. A. (Gordon Alban), 1927–
The valence bond method : theory and practice / G. A. Gallup.
p.    cm.
Includes bibliographical references and index.
ISBN 0 521 80392 6
1. Valence (Theoretical chemistry)    I. Title.
QD469 .G35    2002
541.2′24–dc21                                  2002023398

ISBN-13 978-0-521-80392-2 hardback
ISBN-10 0-521-80392-6 hardback

ISBN-13 978-0-521-02127-2 paperback
ISBN-10 0-521-02127-8 paperback

To my wife Grace, for all her encouragement, and to the memory of our son, Michael, 1956–1995.

# Contents

# Preface

One senses that it is out of style these days to write a book in the sciences all on one's own. Most works coming out today are edited compilations of others' articles collected into chapter-like organization. Perhaps one reason for this is the sheer size of the scientific literature, and the resulting feelings of incompetence engendered, although less honorable reasons are conceivable. Nevertheless, I have attempted this task and submit this book on various aspects of what is called *ab initio* valence bond theory. In it I hope to have made a presentation that is useful for bringing the beginner along as well as presenting material of interest to one who is already a specialist. I have taught quantum mechanics to many students in my career and have come to the conclusion that the beginner frequently confuses the intricacies of mathematical arguments with subtlety. In this book I have not attempted to shy away from intricate presentations, but have worked at removing, insofar as possible, the more subtle ones. One of the ways of doing this is to give good descriptions of simple problems that can show the motivations we have for proceeding as we do with more demanding problems.

This is a book on one sort of model or trial wave function that can be used for molecular calculations of chemical or physical interest. It is in no way a book on the foundations of quantum mechanics – there are many that can be recommended. For the beginner one can still do little better than the books by Pauling and Wilson[1] and Eyring, Walter, and Kimbal[2]. A more recent work is by Levine[3], and for a more "physicsish" presentation the book by Messiah[4] is recommended. These are a little weak on the practice of group theory for which Cotton[5] may serve. A more fundamental work on group theory is by Hammermesh[6]. Some further group theory developments, not to my knowledge in any other book, are in Chapter 5. Some of what we do with the theory of symmetric groups is based fairly heavily on a little book by Rutherford[7].

This is a book on *ab initio* valence bond (VB) theory. There is a vast literature on "valence bond theory" – much of it devoted to semiempirical and qualitative discussions of structure and reactivity of many chemical substances. It is not my purpose to touch upon any of this except occasionally. Rather, I will restrict myself principally to the results and interpretation of the *ab initio* version of the theory. It must be admitted that *ab initio* VB applications are limited to smaller systems, but we shall stick to this more limited goal. Within what practitioners call *ab initio* VB theory there are, in broad terms, two different approaches.

- Calculations in which the orbitals used are restricted to being centered on only one atom of the molecule. They are legitimately called "atomic orbitals". Treatments of this sort may have many configurations involving different orbitals. This approach may be considered a direct descendent of the original Heitler–London work, which is discussed in Chapter 2.
- Calculations in which the orbitals range over two or more atomic centers in the molecule. Although the resulting orbitals are not usually called "molecular orbitals" in this context, there might be some justification in doing so. Within this group of methods there are subcategories that will be addressed in the book. Treatments of this sort usually have relatively few configurations and may be considered descendents of the work of Coulson and Fisher, which is discussed in Chapter 3.

Each of these two approaches has its enthusiasts and its critics. I have attempted an even-handed description of them.

At various places in the text there are suggestions for further study to supplement a discussion or to address a question without a currently known answer. The CRUNCH program package developed by the author and his students is available on the Web for carrying out these studies.[1] This program package was used for all of the examples in the book with the exception of those in Sections 2.2–2.6.

I wish to thank Jeffrey Mills who read large parts of the manuscript and made many useful comments with regard to both style and clarity of presentation. Lastly, I wish to thank all of the students I have had. They did much to contribute to this subject. As time passes, there is nothing like a group of interested students to keep one on one's toes.

Lincoln, Nebraska                                                                                   Gordon A. Gallup
November 2001

---

[1] See http://phy-ggallup.unl.edu/crunch

# List of abbreviations

| | |
|---|---|
| AO | atomic orbital |
| CI | configuration interaction |
| CRUNCH | computational resource for understanding chemistry |
| DZP | double-zeta plus polarization |
| EGSO | eigenvector guided sequential orthogonalization |
| ESE | electronic Schrödinger equation |
| GAMESS | general atomic and molecular electronic structure system |
| GGVB | Goddard's generalized valence bond |
| GUGA | graphical unitary group approach |
| HLSP | Heitler–London–Slater–Pauling |
| LCAO | linear combination of atomic orbitals |
| LMP | least motion path |
| MCVB | multiconfiguration valence bond |
| MO | molecular orbital |
| MOCI | molecular orbital configuration interaction |
| RHF | spin-restricted Hartree–Fock |
| ROHF | spin-restricted open-shell Hartree–Fock |
| SEP | static-exchange potential |
| SCF | self-consistent-field |
| SCVB | spin coupled valence bond |
| UHF | unrestricted Hartree–Fock |
| VB | valence bond |

# Part I

Theory and two-electron systems

# 1

# Introduction

## 1.1 History

In physics and chemistry making a direct calculation to determine the structure or properties of a system is frequently very difficult. Rather, one assumes at the outset an ideal or asymptotic form and then applies adjustments and corrections to make the calculation adhere to what is believed to be a more realistic picture of nature. The practice is no different in molecular structure calculation, but there has developed, in this field, two different "ideals" and two different approaches that proceed from them.

The approach used first, historically, and the one this book is about, is called the valence bond (VB) method today. Heitler and London[8], in their treatment of the $H_2$ molecule, used a trial wave function that was appropriate for two H atoms at long distances and proceeded to use it for all distances. The ideal here is called the "separated atom limit". The results were qualitatively correct, but did not give a particularly accurate value for the dissociation energy of the H−H bond. After the initial work, others made adjustments and corrections that improved the accuracy. This is discussed fully in Chapter 2. A crucial characteristic of the VB method is that the orbitals of different atoms must be considered as nonorthogonal.

The other approach, proposed slightly later by Hund[9] and further developed by Mulliken[10] is usually called the molecular orbital (MO) method. Basically, it views a molecule, particularly a diatomic molecule, in terms of its "united atom limit". That is, $H_2$ is a He atom (not a real one with neutrons in the nucleus) in which the two positive charges are moved from coinciding to the correct distance for the molecule.[1] HF could be viewed as a Ne atom with one proton moved from the nucleus out to the molecular distance, etc. As in the VB case, further adjustments and corrections may be applied to improve accuracy. Although the united atom limit is not often mentioned in work today, its heritage exists in that MOs are universally

---

[1] Although this is impossible to do in practice, we can certainly calculate the process on paper.

considered to be mutually orthogonal. We touch only occasionally upon MO theory in this book.

As formulated by Heitler and London, the original VB method, which was easily extendible to other diatomic molecules, supposed that the atoms making up the molecule were in (high-spin) $S$ states. Heitler and Rumer later extended the theory to polyatomic molecules, but the atomic $S$ state restriction was still, with a few exceptions, imposed. It is in this latter work that the famous Rumer[11] diagrams were introduced. Chemists continue to be intrigued with the possibility of correlating the Rumer diagrams with bonding structures, such as the familiar Kekulé and Dewar bonding pictures for benzene.

Slater and Pauling introduced the idea of using whole atomic configurations rather than $S$ states, although, for carbon, the difference is rather subtle. This, in turn, led to the introduction of hybridization and the maximum overlap criterion for bond formation[1].

Serber[12] and Van Vleck and Sherman[13] continued the analysis and introduced symmetric group arguments to aid in dealing with spin. About the same time the Japanese school involving Yamanouchi and Kotani[14] published analyses of the problem using symmetric group methods.

All of the foregoing work was of necessity fairly qualitative, and only the smallest of molecular systems could be handled. After WWII digital computers became available, and it was possible to test many of the qualitative ideas quantitatively.

In 1949 Coulson and Fisher[15] introduced the idea of nonlocalized orbitals to the VB world. Since that time, suggested schemes have proliferated, all with some connection to the original VB idea. As these ideas developed, the importance of the spin degeneracy problem emerged, and VB methods frequently were described and implemented in this context. We discuss this more fully later.

As this is being written at the beginning of the twenty-first century, even small computers have developed to the point where *ab initio* VB calculations that required "supercomputers" earlier can be carried out in a few minutes or at most a few hours. The development of parallel "supercomputers", made up of many inexpensive personal computer units is only one of the developments that may allow one to carry out ever more extensive *ab initio* VB calculations to look at and interpret molecular structure and reactivity from that unique viewpoint.

## 1.2 Mathematical background

Data on individual atomic systems provided most of the clues physicists used for constructing quantum mechanics. The high spherical symmetry in these cases allows significant simplifications that were of considerable usefulness during times when procedural uncertainties were explored and debated. When the time came to examine the implications of quantum mechanics for molecular structure, it was immediately clear that the lower symmetry, even in diatomic molecules, causes significantly greater difficulties than those for atoms, and nonlinear polyatomic molecules are considerably more difficult still. The mathematical reasons for this are well understood, but it is beyond the scope of this book to pursue these questions. The interested reader may investigate many of the standard works detailing the properties of Lie groups and their applications to physics. There are many useful analytic tools this theory provides for aiding in the solution of partial differential equations, which is the basic mathematical problem we have before us.

### *1.2.1 Schrödinger's equation*

Schrödinger's space equation, which is the starting point of most discussions of molecular structure, is the partial differential equation mentioned above that we must deal with. Again, it is beyond the scope of this book to give even a review of the foundations of quantum mechanics, therefore, we assume Schrödinger's space equation as our starting point. Insofar as we ignore relativistic effects, it describes the energies and interactions that predominate in determining molecular structure. It describes in quantum mechanical terms the kinetic and potential energies of the particles, how they influence the wave function, and how that wave function, in turn, affects the energies. We take up the potential energy term first.

### *Coulomb's law*

Molecules consist of electrons and nuclei; the principal difference between a molecule and an atom is that the latter has only one particle of the nuclear sort. Classical potential theory, which in this case works for quantum mechanics, says that Coulomb's law operates between charged particles. This asserts that the potential energy of a pair of spherical, charged objects is

$$V(|\vec{r}_1 - \vec{r}_2|) = \frac{q_1 q_2}{|\vec{r}_1 - \vec{r}_2|} = \frac{q_1 q_2}{r_{12}}, \tag{1.1}$$

where $q_1$ and $q_2$ are the charges on the two particles, and $r_{12}$ is the scalar distance between them.

### *Units*

A short digression on units is perhaps appropriate here. We shall use either Gaussian units in this book or, much more frequently, Hartree's atomic units. Gaussian units, as far as we are concerned, are identical with the old cgs system of units with the added proviso that charges are measured in unnamed *electrostatic units*, esu. The value of $|e|$ is thus $4.803206808 \times 10^{-10}$ esu. Keeping this number at hand is all that will be required to use Gaussian units in this book.

Hartree's atomic units are usually all we will need. These are obtained by assigning mass, length, and time units so that the mass of the electron, $m_e = 1$, the electronic charge, $|e| = 1$, and Planck's constant, $\hbar = 1$. An upshot of this is that the Bohr radius is also 1. If one needs to compare energies that are calculated in atomic units (hartrees) with measured quantities it is convenient to know that 1 hartree is 27.211396 eV, $6.27508 \times 10^5$ cal/mole, or $2.6254935 \times 10^6$ joule/mole. The reader should be cautioned that one of the most common pitfalls of using atomic units is to forget that the charge on the electron is $-1$. Since equations written in atomic units have no $m_e$s, $e$s, or $\hbar$s in them explicitly, their being all equal to 1, it is easy to lose track of the signs of terms involving the electronic charge. For the moment, however, we continue discussing the potential energy expression in Gaussian units.

### The full potential energy

One of the remarkable features of Coulomb's law when applied to nuclei and electrons is its additivity. The potential energy of an assemblage of particles is just the sum of all the pairwise interactions in the form given in Eq. (1.1). Thus, consider a system with $K$ nuclei, $\alpha = 1, 2, \ldots, K$ having atomic numbers $Z_\alpha$. We also consider the molecule to have $N$ electrons. If the molecule is uncharged as a whole, then $\sum Z_\alpha = N$. We will use lower case Latin letters, $i, j, k, \ldots$, to label electrons and lower case Greek letters, $\alpha, \beta, \gamma, \ldots$, to label nuclei. The full potential energy may then be written

$$V = \sum_{\alpha < \beta} \frac{e^2 Z_\alpha Z_\beta}{r_{\alpha\beta}} - \sum_{i\alpha} \frac{e^2 Z_\alpha}{r_{i\alpha}} + \sum_{i<j} \frac{e^2}{r_{ij}}. \tag{1.2}$$

Many investigations have shown that any deviations from this expression that occur in reality are many orders of magnitude smaller than the sizes of energies we need be concerned with.[2] Thus, we consider this expression to represent exactly that part of the potential energy due to the charges on the particles.

### The kinetic energy

The kinetic energy in the Schrödinger equation is a rather different sort of quantity, being, in fact, a differential operator. In one sense, it is significantly simpler than the potential energy, since the kinetic energy of a particle depends only upon what it is doing, and not on what the other particles are doing. This may be contrasted with the potential energy, which depends not only on the position of the particle in question, but on the positions of all of the other particles, also. For our molecular

---

[2] The first correction to this expression arises because the transmission of the electric field from one particle to another is not instantaneous, but must occur at the speed of light. In electrodynamics this phenomenon is called a *retarded potential*. Casimir and Polder[16] have investigated the consequences of this for quantum mechanics. The effect within distances around $10^{-7}$ cm is completely negligible.

system the kinetic energy operator is

$$T = -\sum_\alpha \frac{\hbar^2}{2M_\alpha} \nabla_\alpha^2 - \sum_i \frac{\hbar^2}{2m_e} \nabla_i^2, \qquad (1.3)$$

where $M_\alpha$ is the mass of the $\alpha^{\text{th}}$ nucleus.

### The differential equation

The Schrödinger equation may now be written symbolically as

$$(T + V)\Psi = E\Psi, \qquad (1.4)$$

where $E$ is the numerical value of the total energy, and $\Psi$ is the wave function. When Eq. (1.4) is solved with the various constraints required by the rules of quantum mechanics, one obtains the total energy and the wave function for the molecule. Other quantities of interest concerning the molecule may subsequently be determined from the wave function.

It is essentially this equation about which Dirac[17] made the famous (or infamous, depending upon your point of view) statement that all of chemistry is reduced to physics by it:

The general theory of quantum mechanics is now almost complete, the imperfections that still remain being in connection with the exact fitting in of the theory with relativity ideas. These give rise to difficulties only when high-speed particles are involved, and are therefore of no importance in the consideration of atomic and molecular structure and ordinary chemical reactions . . . . The underlying physical laws necessary for the mathematical theory of a large part of physics and the whole of chemistry are thus completely known, and the difficulty is only that the exact application of these laws leads to equations much too complicated to be soluble . . . .

To some, with what we might call a practical turn of mind, this seems silly. Our mathematical and computational abilities are not even close to being able to give useful general solutions to it. To those with a more philosophical outlook, it seems significant that, at our present level of understanding, Dirac's statement is apparently true. Therefore, progress made in methods of solving Eq. (1.4) is improving our ability at making predictions from this equation that are useful for answering chemical questions.

### The Born–Oppenheimer approximation

In the early days of quantum mechanics Born and Oppenheimer[18] showed that the energy and motion of the nuclei and electrons could be separated approximately. This was accomplished using a perturbation treatment in which the perturbation parameter is $(m_e/M)^{1/4}$. In actuality, the term "Born–Oppenheimer approximation" is frequently ambiguous. It can refer to two somewhat different theories. The first is the reference above and the other one is found in an appendix of the book by Born and Huang on crystal structure[19]. In the latter treatment, it is assumed, based upon physical arguments, that the wave function of Eq. (1.4) may be written as the product of two other functions

$$\Psi(\vec{r}_i, \vec{r}_\alpha) = \phi(\vec{r}_\alpha)\psi(\vec{r}_i, \vec{r}_\alpha), \tag{1.5}$$

where the nuclear positions $\vec{r}_\alpha$ given in $\psi$ are parameters rather than variables in the normal sense. The $\phi$ is the actual wave function for nuclear motion and will not concern us at all in this book. If Eq. (1.5) is substituted into Eq. (1.4), various terms are collected, and small quantities dropped, we obtain what is frequently called the Schrödinger equation for the electrons using the Born–Oppenheimer approximation

$$-\frac{\hbar^2}{2m_e}\sum_i \nabla_i^2\psi + V\psi = E(\vec{r}_\alpha)\psi, \tag{1.6}$$

where we have explicitly observed the dependence of the energy on the nuclear positions by writing it as $E(\vec{r}_\alpha)$. Equation (1.6) might better be termed the Schrödinger equation for the electrons using the *adiabatic* approximation[20]. Of course, the only difference between this and Eq. (1.4) is the presence of the nuclear kinetic energy in the latter. A heuristic way of looking at Eq. (1.6) is to observe that it would arise if the masses of the nuclei all passed to infinity, i.e., the nuclei become stationary. Although a physically useful viewpoint, the actual validity of such a procedure requires some discussion, which we, however, do not give.

We now go farther, introducing atomic units and rearranging Eq. (1.6) slightly,

$$-\frac{1}{2}\sum_i \nabla_i^2\psi - \sum_{i\alpha}\frac{Z_\alpha}{r_{i\alpha}}\psi + \sum_{i<j}\frac{1}{r_{ij}}\psi + \sum_{\alpha<\beta}\frac{Z_\alpha Z_\beta}{r_{\alpha\beta}}\psi = E_e\psi. \tag{1.7}$$

This is the equation with which we must deal. We will refer to it so frequently, it will be convenient to have a brief name for it. It is the *electronic Schrödinger equation*, and we refer to it as the ESE. Solutions to it of varying accuracy have been calculated since the early days of quantum mechanics. Today, there exist computer programs both commercial and in the public domain that will carry out calculations to produce approximate solutions to the ESE. Indeed, a program of this sort is available from the author through the Internet.[3] Although not as large as some of the others available, it will do many of the things the bigger programs will do, as well as a couple of things they do not: in particular, this program will do VB calculations of the sort we discuss in this book.

---

[3] The CRUNCH program, http://phy-ggallup.unl.edu/crunch/

## 1.3 The variation theorem

### 1.3.1 General variation functions

If we write the sum of the kinetic and potential energy operators as the Hamiltonian operator $T + V = H$, the ESE may be written as

$$H\Psi = E\Psi. \qquad (1.8)$$

One of the remarkable results of quantum mechanics is the variation theorem, which states that

$$W = \frac{\langle \Psi | H | \Psi \rangle}{\langle \Psi | \Psi \rangle} \geq E_0, \qquad (1.9)$$

where $E_0$ is the lowest allowed eigenvalue for the system. The fraction in Eq. (1.9) is frequently called the *Rayleigh quotient*. The basic use of this result is quite simple. One uses arguments based on similarity, intuition, guess-work, or whatever, to devise a suitable function for $\Psi$. Using Eq. (1.9) then necessarily gives us an upper bound to the true lowest energy, and, if we have been clever or lucky, the upper bound is a good approximation to the lowest energy. The most common way we use this is to construct a trial function, $\Psi$, that has a number of parameters in it. The quantity, $W$, in Eq. (1.9) is then a function of these parameters, and a minimization of $W$ with respect to the parameters gives the best result possible within the limitations of the choice for $\Psi$. We will use this scheme in a number of discussions throughout the book.

### 1.3.2 Linear variation functions

A trial variation function that has linear variation parameters only is an important special case, since it allows an analysis giving a systematic improvement on the lowest upper bound as well as upper bounds for excited states. We shall assume that $\phi_1, \phi_2, \ldots$, represents a complete, normalized (but not necessarily orthogonal) set of functions for expanding the exact eigensolutions to the ESE. Thus we write

$$\Psi = \sum_{i=1}^{\infty} \phi_i C_i, \qquad (1.10)$$

where the $C_i$ are the variation parameters. Substituting into Eq. (1.9) we obtain

$$W = \frac{\sum_{ij} H_{ij} C_i^* C_j}{\sum_{ij} S_{ij} C_i^* C_j}, \qquad (1.11)$$

where

$$H_{ij} = \langle \phi_i | H | \phi_j \rangle, \qquad (1.12)$$
$$S_{ij} = \langle \phi_i | \phi_j \rangle. \qquad (1.13)$$

We differentiate $W$ with respect to the $C_i^*$s and set the results to zero to find the minimum, obtaining an equation for each $C_i^*$,

$$\sum_j (H_{ij} - W S_{ij}) C_j = 0; \qquad i = 1, 2, \ldots . \tag{1.14}$$

In deriving this we have used the properties of the integrals $H_{ij} = H_{ji}^*$ and a similar result for $S_{ij}$. Equation (1.14) is discussed in all elementary textbooks wherein it is shown that a $C_j \neq 0$ solution exists only if the $W$ has a specific set of values. It is sometimes called the *generalized eigenvalue problem* to distinguish from the case when $S$ is the identity matrix. We wish to pursue further information about the $W$s here.

Let us consider a variation function where we have chosen $n$ of the functions, $\phi_i$. We will then show that the eigenvalues of the $n$-function problem divide, i.e., occur between, the eigenvalues of the $(n + 1)$-function problem. In making this analysis we use an extension of the methods given by Brillouin[21] and MacDonald[22].

Having chosen $n$ of the $\phi$ functions to start, we obtain an equation like Eq. (1.14), but with only $n \times n$ matrices and $n$ terms,

$$\sum_{j=1}^n \left( H_{ij} - W^{(n)} S_{ij} \right) C_j^{(n)} = 0; \qquad i = 1, 2, \ldots, n. \tag{1.15}$$

It is well known that sets of linear equations like Eq. (1.15) will possess nonzero solutions for the $C_j^{(n)}$s only if the matrix of coefficients has a rank less than $n$. This is another way of saying that the determinant of the matrix is zero, so we have

$$\left| H - W^{(n)} S \right| = 0. \tag{1.16}$$

When expanded out, the determinant is a polynomial of degree $n$ in the variable $W^{(n)}$, and it has $n$ real roots if $H$ and $S$ are both Hermitian matrices, and $S$ is positive definite. Indeed, if $S$ were not positive definite, this would signal that the basis functions were not all linearly independent, and that the basis was defective. If $W^{(n)}$ takes on one of the roots of Eq. (1.16) the matrix $H - W^{(n)} S$ is of rank $n - 1$ or less, and its rows are linearly dependent. There is thus at least one more nonzero vector with components $C_j^{(n)}$ that can be orthogonal to all of the rows. This is the solution we want.

It is useful to give a matrix solution to this problem. We affix a superscript $^{(n)}$ to emphasize that we are discussing a matrix solution for $n$ basis functions. Since $S^{(n)}$ is Hermitian, it may be diagonalized by a unitary matrix, $T = (T^\dagger)^{-1}$

$$T^\dagger S^{(n)} T = s^{(n)} = \operatorname{diag}\left(s_1^{(n)}, s_2^{(n)}, \ldots, s_n^{(n)}\right), \tag{1.17}$$

where the diagonal elements of $s^{(n)}$ are all real and positive, because of the Hermitian and positive definite character of the overlap matrix. We may construct the inverse square root of $s^{(n)}$, and, clearly, we obtain

$$\left[T\left(s^{(n)}\right)^{-1/2}\right]^{\dagger} S^{(n)} T\left(s^{(n)}\right)^{-1/2} = I. \tag{1.18}$$

We subject $H^{(n)}$ to the same transformation and obtain

$$\left[T\left(s^{(n)}\right)^{-1/2}\right]^{\dagger} H^{(n)} T\left(s^{(n)}\right)^{-1/2} = \bar{H}^{(n)}, \tag{1.19}$$

which is also Hermitian and may be diagonalized by a unitary matrix, $U$. Combining the various transformations, we obtain

$$V^{\dagger} H^{(n)} V = h^{(n)} = \operatorname{diag}\left(h_1^{(n)}, h_2^{(n)}, \ldots, h_n^{(n)}\right), \tag{1.20}$$
$$V^{\dagger} S^{(n)} V = I, \tag{1.21}$$
$$V = T\left(s^{(n)}\right)^{-1/2} U. \tag{1.22}$$

We may now combine these matrices to obtain the null matrix

$$V^{\dagger} H^{(n)} V - V^{\dagger} S^{(n)} V h^{(n)} = 0, \tag{1.23}$$

and multiplying this on the left by $(V^{\dagger})^{-1} = U(s^{(n)})^{1/2} T$ gives

$$H^{(n)} V - S^{(n)} V h^{(n)} = 0. \tag{1.24}$$

If we write out the $k^{\text{th}}$ column of this last equation, we have

$$\sum_{j=1}^{n} \left(H_{ij}^{(n)} - h_k^{(n)} S_{ij}^{(n)}\right) V_{jk} = 0; \qquad i = 1, 2, \ldots, n. \tag{1.25}$$

When this is compared with Eq. (1.15) we see that we have solved our problem, if $C^{(n)}$ is the $k^{\text{th}}$ column of $V$ and $W^{(n)}$ is the $k^{\text{th}}$ diagonal element of $h^{(n)}$. Thus the diagonal elements of $h^{(n)}$ are the roots of the determinantal equation Eq. (1.16).

Now consider the variation problem with $n + 1$ functions where we have added another of the basis functions to the set. We now have the matrices $H^{(n+1)}$ and $S^{(n+1)}$, and the new determinantal equation

$$\left|H^{(n+1)} - W^{(n+1)} S^{(n+1)}\right| = 0. \tag{1.26}$$

We may subject this to a transformation by the $(n + 1) \times (n + 1)$ matrix

$$\bar{V} = \begin{bmatrix} V & 0 \\ 0 & 1 \end{bmatrix}, \tag{1.27}$$

and $H^{(n+1)}$ and $S^{(n+1)}$ are modified to

$$
\bar{V}^{\dagger} H^{(n+1)} \bar{V} = \bar{H}^{(n+1)} =
\begin{bmatrix}
h_1^{(n)} & 0 & \cdots & \bar{H}_{1\,n+1}^{(n+1)} \\
0 & h_2^{(n)} & \cdots & \bar{H}_{2\,n+1}^{(n+1)} \\
\vdots & \vdots & \ddots & \vdots \\
\bar{H}_{n+1\,1}^{(n+1)} & \bar{H}_{n+1\,2}^{(n+1)} & \cdots & H_{n+1\,n+1}^{n+1}
\end{bmatrix}
\tag{1.28}
$$

and

$$
\bar{V}^{\dagger} S^{(n+1)} \bar{V} = \bar{S}^{(n+1)} =
\begin{bmatrix}
1 & 0 & \cdots & \bar{S}_{1\,n+1}^{(n+1)} \\
0 & 1 & \cdots & \bar{S}_{2\,n+1}^{(n+1)} \\
\vdots & \vdots & \ddots & \vdots \\
\bar{S}_{n+1\,1}^{(n+1)} & \bar{S}_{n+1\,2}^{(n+1)} & \cdots & 1
\end{bmatrix} .
\tag{1.29}
$$

Thus Eq. (1.26) becomes

$$
0 =
\begin{vmatrix}
h_1^{(n)} - W^{(n+1)} & 0 & \cdots & \bar{H}_{1\,n+1}^{(n+1)} - W^{(n+1)} \bar{S}_{1\,n+1}^{(n+1)} \\
0 & h_2^{(n)} - W^{(n+1)} & \cdots & \bar{H}_{2\,n+1}^{(n+1)} - W^{(n+1)} \bar{S}_{2\,n+1}^{(n+1)} \\
\vdots & \vdots & \ddots & \vdots \\
\bar{H}_{n+1\,1}^{(n+1)} - W^{(n+1)} \bar{S}_{n+1\,1}^{(n+1)} & \bar{H}_{n+1\,2}^{(n+1)} - W^{(n+1)} \bar{S}_{n+1\,2}^{(n+1)} & \cdots & H_{n+1\,n+1}^{n+1} - W^{(n+1)}
\end{vmatrix} .
$$

$$\tag{1.30}$$

We modify the determinant in Eq. (1.30) by using column operations. Multiply the $i^{\text{th}}$ column by

$$
\frac{\bar{H}_{i\,n+1}^{(n+1)} - W^{(n+1)} \bar{S}_{i\,n+1}^{(n+1)}}{h_i^{(n)} - W^{(n+1)}}
$$

and subtract it from the $(n+1)^{\text{th}}$ column. This is seen to cancel the $i^{\text{th}}$ row element in the last column. Performing this action for each of the first $n$ columns, the determinant is converted to lower triangular form, and its value is just the product of the diagonal elements,

$$
0 = D^{(n+1)}\left(W^{(n+1)}\right)
$$

$$
= \prod_{i=1}^{n} \left[ h_i^{(n)} - W^{(n+1)} \right]
$$

$$
\times \left[ \bar{H}_{n+1\,n+1}^{(n)} - W^{(n+1)} - \sum_{i=1}^{n} \frac{\left| \bar{H}_{i\,n+1}^{(n+1)} - W^{(n+1)} \bar{S}_{i\,n+1}^{(n+1)} \right|^2}{h_i^{(n)} - W^{(n+1)}} \right] .
\tag{1.31}
$$

Examination shows that $D^{(n+1)}(W^{(n+1)})$ is a polynomial in $W^{(n+1)}$ of degree $n+1$, as it should be.

We note that none of the $h_i^{(n)}$ are normally roots of $D^{(n+1)}$,

$$\lim_{W^{(n+1)} \to h_i^{(n)}} D^{(n+1)} = \prod_{j \neq i} [h_j^{(n)} - h_i^{(n)}] |\bar{H}_{i\,n+1}^{(n+1)} - h_i^{(n)} \bar{S}_{i\,n+1}^{(n+1)}|^2, \quad (1.32)$$

and would be only if the $h_i^{(n)}$ were degenerate or the second factor $|\cdots|^2$ were zero.[4]

Thus, $D^{(n+1)}$ is zero when the second $[\cdots]$ factor of Eq. (1.31) is zero,

$$\bar{H}_{n+1\,n+1}^{(n+1)} - W^{(n+1)} = \sum_{i=1}^{n} \frac{|\bar{H}_{i\,n+1}^{(n+1)} - W^{(n+1)} \bar{S}_{i\,n+1}^{(n+1)}|^2}{h_i^{(n)} - W^{(n+1)}}. \quad (1.33)$$

It is most useful to consider the solution of Eq. (1.33) graphically by plotting both the right and left hand sides versus $W^{(n+1)}$ on the same graph and determining where the two curves cross. For this purpose let us suppose that $n = 4$, and we consider the right hand side. It will have poles on the real axis at each of the $h_i^{(4)}$. When $W^{(5)}$ becomes large in either the positive or negative direction the right hand side asymptotically approaches the line

$$y = \sum_{i=1}^{4} (\bar{H}_{i\,5}^* \bar{S}_{i\,5} + \bar{H}_{i\,5} \bar{S}_{i\,5}^* - W^{(5)} |\bar{S}_{i\,5}^{(5)}|^2).$$

It is easily seen that the determinant of $\bar{S}$ is

$$|\bar{S}| = 1 - \sum_{i=1}^{4} |\bar{S}_{i\,5}^{(5)}|^2 > 0, \quad (1.34)$$

and, if equal to zero, $S$ would not be positive definite, a circumstance that would happen only if our basis were linearly dependent. Thus, the asymptotic line of the right hand side has a slope between 0 and $-45°$. We see this in Fig. 1.1. The left hand side of Eq. (1.33) is, on the other hand, just a straight line of exactly $-45°$ slope and a $W^{(5)}$ intercept of $\bar{H}_{55}^{(5)}$. This is also shown in Fig. 1.1. The important point we note is that the right hand side of Eq. (1.33) has five branches that intersect the left hand line in five places, and we thus obtain five roots. The vertical dotted lines in Fig. 1.1 are the values of the $h_i^{(4)}$, and we see there is one of these between each pair of roots for the five-function problem. A little reflection will indicate that this important fact is true for any $n$, not just the special case plotted in Fig. 1.1.

---

[4] We shall suppose neither of these possibilities occurs, and in practice neither is likely in the absence of symmetry. If there is symmetry present that can produce degeneracy or zero factors of the $[\cdots]^2$ sort, we assume that symmetry factorization has been applied and that all functions we are working with are within one of the closed symmetry subspaces of the problem.

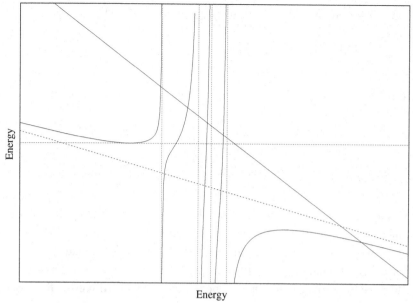

Energy

Figure 1.1. The relationship between the roots for $n = 4$ (the abscissa intercepts of the vertical dotted lines) and $n = 5$ (abscissas of intersections of solid lines with solid curves) shown graphically.

The upshot of these considerations is that a series of matrix solutions of the variation problem, where we add one new function at a time to the basis, will result in a series of eigenvalues in a pattern similar to that shown schematically in Fig. 1.2, and that the order of adding the functions is immaterial. Since we suppose that our ultimate basis ($n \rightarrow \infty$) is complete, each of the eigenvalues will become exact as we pass to an infinite basis, and we see that the sequence of $n$-basis solutions converges to the correct answer from above. The rate of convergence at various levels will certainly depend upon the order in which the basis functions are added, but not the ultimate value.

### 1.3.3 A 2 × 2 generalized eigenvalue problem

The generalized eigenvalue problem is unfortunately considerably more complicated than its regular counterpart when $S = I$. There are possibilities for accidental cases when basis functions apparently should mix, but they do not. We can give a simple example of this for a $2 \times 2$ system. Assume we have the pair of matrices

$$H = \begin{bmatrix} A & B \\ B & C \end{bmatrix} \qquad (1.35)$$

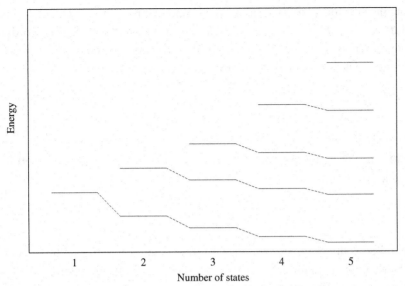

Figure 1.2. A qualitative graph showing schematically the interleaving of the eigenvalues for a series of linear variation problems for $n = 1, \ldots, 5$. The ordinate is energy.

and

$$S = \begin{bmatrix} 1 & s \\ s & 1 \end{bmatrix}, \tag{1.36}$$

where we assume for the argument that $s > 0$. We form the matrix $H'$

$$H' = H - \frac{A+C}{2}S,$$

$$= \begin{bmatrix} a & b \\ b & -a \end{bmatrix}, \tag{1.37}$$

where

$$a = A - \frac{A+C}{2} \tag{1.38}$$

and

$$b = B - \frac{A+C}{2}s. \tag{1.39}$$

It is not difficult to show that the eigenvectors of $H'$ are the same as those of $H$.

Our generalized eigenvalue problem thus depends upon three parameters, $a$, $b$, and $s$. Denoting the eigenvalue by $W$ and solving the quadratic equation, we obtain

$$W = -\frac{sb}{(1-s^2)} \pm \frac{\sqrt{a^2(1-s^2) + b^2}}{(1-s^2)}. \tag{1.40}$$

We note the possibility of an accident that cannot happen if $s = 0$ and $b \neq 0$: Should $b = \pm as$, one of the two values of $W$ is either $\pm a$, and one of the two diagonal elements of $H'$ is unchanged.[5] Let us for definiteness assume that $b = as$ and it is $a$ we obtain. Then, clearly the vector $C_1$ we obtain is

$$\begin{bmatrix} 1 \\ 0 \end{bmatrix},$$

and there is no mixing between the states from the application of the variation theorem. The other eigenvector is simply determined because it must be orthogonal to $C_1$, and we obtain

$$C_2 = \begin{bmatrix} -s/\sqrt{1-s^2} \\ 1/\sqrt{1-s^2} \end{bmatrix},$$

so the other state is mixed. It must normally be assumed that this accident is rare in practical calculations. Solving the generalized eigenvalue problem results in a nonorthogonal basis changing both directions and internal angles to become orthogonal. Thus one basis function could get "stuck" in the process. This should be contrasted with the case when $S = I$, in which basis functions are unchanged only if the matrix was originally already diagonal with respect to them.

We do not discuss it, but there is an $n \times n$ version of this complication. If there is no degeneracy, one of the diagonal elements of the $H$-matrix may be unchanged in going to the eigenvalues, and the eigenvector associated with it is $[0, \ldots, 0, 1, 0, \ldots, 0]^\dagger$.

## 1.4 Weights of nonorthogonal functions

The probability interpretation of the wave function in quantum mechanics obtained by forming the square of its magnitude leads naturally to a simple idea for the weights of constituent parts of the wave function when it is written as a linear combination of orthonormal functions. Thus, if

$$\Psi = \sum_i \psi_i C_i, \tag{1.41}$$

and $\langle \psi_i | \psi_j \rangle = \delta_{ij}$, normalization of $\Psi$ requires

$$\sum_i |C_i|^2 = 1. \tag{1.42}$$

If, also, each of the $\psi_i$ has a certain physical interpretation or significance, then one says the wave function $\Psi$, or the state represented by it, consists of a fraction

---

[5] NB We assumed this not to happen in our discussion above of the convergence in the linear variation problem.

$|C_i|^2$ of the state represented by $\psi_i$. One also says that the *weight*, $w_i$ of $\psi_i$ in $\Psi$ is $w_i = |C_i|^2$.

No such simple result is available for nonorthogonal bases, such as our VB functions, because, although they are normalized, they are not mutually orthogonal. Thus, instead of Eq. (1.42), we would have

$$\sum_{ij} C_i^* C_j S_{ij} = 1, \qquad (1.43)$$

if the $\psi_i$ were not orthonormal. In fact, at first glance orthogonalizing them would mix together characteristics that one might wish to consider separately in determining weights. In the author's opinion, there has not yet been devised a completely satisfactory solution to this problem. In the following paragraphs we mention some suggestions that have been made and, in addition, present yet another way of attempting to resolve this problem.

In Section 2.8 we discuss some simple functions used to represent the $H_2$ molecule. We choose one involving six basis functions to illustrate the various methods. The overlap matrix for the basis is

$$\begin{bmatrix} 1.000\,000 & & & & & \\ 0.962\,004 & 1.000\,000 & & & & \\ 0.137\,187 & 0.181\,541 & 1.000\,000 & & & \\ -0.254\,383 & -0.336\,628 & 0.141\,789 & 1.000\,000 & & \\ 0.181\,541 & 0.137\,187 & 0.925\,640 & 0.251\,156 & 1.000\,000 & \\ 0.336\,628 & 0.254\,383 & -0.251\,156 & -0.788\,501 & -0.141\,789 & 1.000\,000 \end{bmatrix},$$

and the eigenvector we analyze is

$$\begin{bmatrix} 0.283\,129 \\ 0.711\,721 \\ 0.013\,795 \\ -0.038\,111 \\ -0.233\,374 \\ 0.017\,825 \end{bmatrix}. \qquad (1.44)$$

$S$ is to be filled out, of course, so that it is symmetric. The particular chemical or physical significance of the basis functions need not concern us here.

The methods below giving sets of weights fall into one of two classes: those that involve no orthogonalization and those that do. We take up the former group first.

Table 1.1. *Weights for nonorthogonal basis functions*
*by various methods.*

| Chirgwin–Coulson | Inverse-overlap | Symmetric orthogon. | EGSO[a] |
|---|---|---|---|
| 0.266 999 | 0.106 151 | 0.501 707 | 0.004 998 |
| 0.691 753 | 0.670 769 | 0.508 663 | 0.944 675 |
| −0.000 607 | 0.000 741 | 0.002 520 | 0.000 007 |
| 0.016 022 | 0.008 327 | 0.042 909 | 0.002 316 |
| 0.019 525 | 0.212 190 | 0.051 580 | 0.047 994 |
| 0.006 307 | 0.001 822 | 0.000 065 | 0.000 010 |

[a] EGSO = eigenvector guided sequential orthogonalization.

### 1.4.1 Weights without orthogonalization

*The method of Chirgwin and Coulson*

These workers[23] suggest that one use

$$w_i = C_i^* \sum_j S_{ij} C_j, \qquad (1.45)$$

although, admittedly, they proposed it only in cases where the quantities were real. As written, this $w_i$ is not guaranteed even to be real, and when the $C_i$ and $S_{ij}$ are real, it is not guaranteed to be positive. Nevertheless, in simple cases it can give some idea for weights. We show the results of applying this method to the eigenvector and overlap matrix in Table 1.1 above. We see that the relative weights of basis functions 2 and 1 are fairly large and the others are quite small.

*Inverse overlap weights*

Norbeck and the author[24] suggested that in cases where there is overlap, the basis functions each can be considered to have a unique portion. The "length" of this may be shown to be equal to the reciprocal of the diagonal of the $S^{-1}$ matrix corresponding to the basis function in question. Thus, if a basis function has a unique portion of very short length, a large coefficient for it means little. This suggests that a set of *relative* weights could be obtained from

$$w_i \propto |C_i|^2 / (S^{-1})_{ii}, \qquad (1.46)$$

where these $w_i$ do not generally sum to 1. As implemented, these weights are renormalized so that they do sum to 1 to provide convenient fractions or percentages. This is an awkward feature of this method and makes it behave nonlinearly in some contexts. Although these first two methods agree as to the most important basis function they transpose the next two in importance.

### 1.4.2 Weights requiring orthogonalization

We emphasize that here we are speaking of orthogonalizing the VB basis not the underlying atomic orbitals (AOs). This can be accomplished by a transformation of the overlap matrix to convert it to the identity

$$N^\dagger S N = I. \tag{1.47}$$

Investigation shows that $N$ is far from unique. Indeed, if $N$ satisfies Eq. (1.47), $NU$ will also work, where $U$ is any unitary matrix. A possible candidate for $N$ is shown in Eq. (1.18). If we put restrictions on $N$, the result can be made unique. If $N$ is forced to be upper triangular, one obtains the classical *Schmidt orthogonalization* of the basis. The transformation of Eq. (1.18), as it stands, is frequently called the *canonical orthogonalization* of the basis. Once the basis is orthogonalized the weights are easily determined in the normal sense as

$$w_i = \left| \sum_j (N^{-1})_{ij} C_j \right|^2, \tag{1.48}$$

and, of course, they sum to 1 exactly without modification.

#### Symmetric orthogonalization

Löwdin[25] suggested that one find the orthonormal set of functions that most closely approximates the original nonorthogonal set in the least squares sense and use these to determine the weights of various basis functions. An analysis shows that the appropriate transformation in the notation of Eq. (1.18) is

$$N = T\left(s^{(n)}\right)^{-1/2} T^\dagger = S^{-1/2} = (S^{-1/2})^\dagger, \tag{1.49}$$

which is seen to be the inverse of one of the square roots of the overlap matrix and Hermitian (symmetric, if real). Because of this symmetry, using the $N$ of Eq. (1.49) is frequently called a *symmetric orthogonalization*. This translates easily into the set of weights

$$w_i = \left| \sum_j (S^{1/2})_{ij} C_j \right|^2, \tag{1.50}$$

which sums to 1 without modification. These are also shown in Table 1.1. We now see weights that are considerably different from those in the first two columns. $w_1$ and $w_2$ are nearly equal, with $w_2$ only slightly larger. This is a direct result of the relatively large value of $S_{12}$ in the overlap matrix, but, indirectly, we note that the hypothesis behind the symmetric orthogonalization can be faulty. A least squares problem like that resulting in this orthogonalization method, in principle, always has an answer, but that gives no guarantee at all that the functions produced really are close to the original ones. That is really the basic difficulty. Only if the overlap matrix were, in some sense, close to the identity would this method be expected to yield useful results.

*An eigenvector guided sequential orthogonalization (EGSO)*

As promised, with this book we introduce another suggestion for determining weights in VB functions. Let us go back to one of the ideas behind inverse overlap weights and apply it differently. The existence of nonzero overlaps between different basis functions suggests that some "parts" of basis functions are duplicated in the sum making up the total wave function. At the same time, consider function 2 (the second entry in the eigenvector (1.44)). The eigenvector was determined using linear variation functions, and clearly, there is something about function 2 that the variation theorem likes, it has the largest (in magnitude) coefficient. Therefore, we take all of that function in our orthogonalization, and, using a procedure analogous to the Schmidt procedure, orthogonalize all of the remaining functions of the basis to it. This produces a new set of $C$s, and we can carry out the process again with the largest remaining coefficient. We thus have a stepwise procedure to orthogonalize the basis. Except for the order of choice of functions, this is just a Schmidt orthogonalization, which normally, however, involves an arbitrary or preset ordering.

Comparing these weights to the others in Table 1.1 we note that there is now one truly dominant weight and the others are quite small. Function 2 is really a considerable portion of the total function at 94.5%. Of the remaining, only function 5 at 4.8% has any size. It is interesting that the two methods using somewhat the same idea predict the same two functions to be dominant.

If we apply this procedure to a different state, there will be a different ordering, in general, but this is expected. The orthogonalization in this procedure is not designed to generate a basis for general use, but is merely a device to separate characteristics of basis functions into noninteracting pieces that allows us to determine a set of weights. Different eigenvalues, i.e., different states, may well be quite different in this regard.

We now outline the procedure in more detail. Deferring the question of ordering until later, let us assume we have found an upper triangular transformation matrix, $N_k$, that converts $S$ as follows:

$$(N_k)^\dagger S N_k = \begin{bmatrix} I_k & 0 \\ 0 & S_{n-k} \end{bmatrix}, \tag{1.51}$$

where $I_k$ is a $k \times k$ identity, and we have determined $k$ of the orthogonalized weights. We show how to determine $N_{k+1}$ from $N_k$.

Working only with the lower right $(n-k) \times (n-k)$ corner of the matrices, we observe that $S_{n-k}$ in Eq. (1.51) is just the overlap matrix for the unreduced portion of the basis and is, in particular, Hermitian, positive definite, and with diagonal elements equal to 1. We write it in partitioned form as

$$S_{n-k} = \begin{bmatrix} 1 & s \\ s^\dagger & S' \end{bmatrix},$$ (1.52)

where $[1 \quad s]$ is the first row of the matrix. Let $M_{n-k}$ be an upper triangular matrix partitioned similarly,

$$M_{n-k} = \begin{bmatrix} 1 & q \\ 0 & B \end{bmatrix},$$ (1.53)

and we determine $q$ and $B$ so that

$$(M_{n-k})^\dagger S_{n-k} M_{n-k} = \begin{bmatrix} 1 & q + sB \\ (q+sB)^\dagger & B^\dagger(S' - s^\dagger s)B \end{bmatrix},$$ (1.54)

$$= \begin{bmatrix} 1 & 0 \\ 0 & S_{n-k-1} \end{bmatrix},$$ (1.55)

where these equations may be satisfied with $B$ the diagonal matrix

$$B = \text{diag}\left((1 - s_1^2)^{-1/2} (1 - s_2^2)^{-1/2} \cdots\right)$$ (1.56)

and

$$q = -sB.$$ (1.57)

The inverse of $M_{n-k}$ is easily determined:

$$(M_{n-k})^{-1} = \begin{bmatrix} 1 & s \\ 0 & B^{-1} \end{bmatrix},$$ (1.58)

and, thus, $N_{k+1} = N_k Q_k$, where

$$Q_k = \begin{bmatrix} I_k & 0 \\ 0 & M_{n-k} \end{bmatrix}.$$ (1.59)

The unreduced portion of the problem is now transformed as follows:

$$(C_{n-k})^\dagger S_{n-k} C_{n-k} = [(M_{n-k})^{-1} C_{n-k}]^\dagger (M_{n-k})^\dagger S_{n-k} M_{n-k}[(M_{n-k})^{-1} C_{n-k}].$$ (1.60)

Writing

$$C_{n-k} = \begin{bmatrix} C_1 \\ C' \end{bmatrix},$$ (1.61)

we have

$$[(M_{n-k})^{-1} C_{n-k}] = \begin{bmatrix} C_1 + sC' \\ B^{-1}C' \end{bmatrix},$$ (1.62)

$$= \begin{bmatrix} C_1 + sC' \\ C_{n-k-1} \end{bmatrix}.$$ (1.63)

Putting these together, we arrive at the total $N$ as $Q_1 Q_2 Q_3 \cdots Q_{n-1}$.

What we have done so far is, of course, no different from a standard top-down Schmidt orthogonalization. We wish, however, to guide the ordering with the eigenvector. This we accomplish by inserting before each $Q_k$ a binary permutation matrix $P_k$ that puts in the top position the $C_1 + sC'$ from Eq. (1.63) that is largest in magnitude. Our actual transformation matrix is

$$N = P_1 Q_1 P_2 Q_2 \cdots P_{n-1} Q_{n-1}. \tag{1.64}$$

Then the weights are simply as given (for basis functions in a different order) by Eq. (1.48). We observe that choosing $C_1 + sC'$ as the test quantity whose magnitude is maximized is the same as choosing the remaining basis function from the unreduced set that at each stage gives the greatest contribution to the total wave function.

There are situations in which we would need to modify this procedure for the results to make sense. Where symmetry dictates that two or more basis functions should have equal contributions, the above algorithm could destroy this equality. In these cases some modification of the procedure is required, but we do not need this extension for the applications of the EGSO weights found in this book.

# 2

# H$_2$ and localized orbitals

## 2.1 The separation of spin and space variables

One of the pedagogically unfortunate aspects of quantum mechanics is the complexity that arises in the interaction of electron spin with the Pauli exclusion principle as soon as there are more than two electrons. In general, since the ESE does not even contain any spin operators, the total spin operator must commute with it, and, thus, the total spin of a system of any size is conserved at this level of approximation. The corresponding solution to the ESE must reflect this. In addition, the total electronic wave function must also be antisymmetric in the interchange of any pair of space-spin coordinates, and the interaction of these two requirements has a subtle influence on the energies that has no counterpart in classical systems.

### 2.1.1 The spin functions

When there are only two electrons the analysis is much simplified. Even quite elementary textbooks discuss two-electron systems. The simplicity is a consequence of the general nature of what is called the *spin-degeneracy problem*, which we describe in Chapters 4 and 5. For now we write the total solution for the ESE $\Psi(1, 2)$, where the labels 1 and 2 refer to the coordinates (space and spin) of the two electrons. Since the ESE has no reference at all to spin, $\Psi(1, 2)$ may be factored into separate spatial and spin functions. For two electrons one has the familiar result that the spin functions are of either the singlet or triplet type,

$$^1\phi_0 = \left[\eta_{1/2}(1)\eta_{-1/2}(2) - \eta_{-1/2}(1)\eta_{1/2}(2)\right]\big/\sqrt{2}, \tag{2.1}$$

$$^3\phi_1 = \eta_{1/2}(1)\eta_{1/2}(2), \tag{2.2}$$

$$^3\phi_0 = \left[\eta_{1/2}(1)\eta_{-1/2}(2) + \eta_{-1/2}(1)\eta_{1/2}(2)\right]\big/\sqrt{2}, \tag{2.3}$$

$$^3\phi_{-1} = \eta_{-1/2}(1)\eta_{-1/2}(2), \tag{2.4}$$

where on the $\phi$ the anterior superscript indicates the multiplicity and the posterior subscript indicates the $m_s$ value. The $\eta_{\pm 1/2}$ are the individual electron spin functions.

If we let $P_{ij}$ represent an operator that interchanges all of the coordinates of the $i^{\text{th}}$ and $j^{\text{th}}$ particles in the function to which it is applied, we see that

$$P_{12}\, {}^1\phi_0 = -{}^1\phi_0, \tag{2.5}$$

$$P_{12}\, {}^3\phi_{m_s} = {}^3\phi_{m_s}; \tag{2.6}$$

thus, the singlet spin function is antisymmetric and the triplet functions are symmetric with respect to interchange of the two sets of coordinates.

### 2.1.2 The spatial functions

The Pauli exclusion principle requires that the total wave function for electrons (fermions) have the property

$$P_{12}\Psi(1, 2) = \Psi(2, 1) = -\Psi(1, 2), \tag{2.7}$$

but absence of spin in the ESE requires

$${}^{(2S+1)}\Psi_{m_s}(1, 2) = {}^{(2S+1)}\psi(1, 2) \times {}^{(2S+1)}\phi_{m_s}(1, 2), \tag{2.8}$$

and it is not hard to see that the overall antisymmetry requires that the spatial function $\psi$ have behavior opposite to that of $\phi$ in all cases. We emphasize that it is not an oversight to attach no $m_s$ label to $\psi$ in Eq. (2.8). An important principle in quantum mechanics, known as the Wigner–Eckart theorem, requires the spatial part of the wave function to be independent of $m_s$ for a given $S$.

Thus the singlet spatial function is symmetric and the triplet one antisymmetric. If we use the variation theorem to obtain an approximate solution to the ESE requiring symmetry as a subsidiary condition, we are dealing with the singlet state for two electrons. Alternatively, antisymmetry, as a subsidiary condition, yields the triplet state.

We must now see how to obtain useful solutions to the ESE that satisfy these conditions.

## 2.2 The AO approximation

The only uncharged molecule with two electrons is $H_2$, and we will consider this molecule for a while. The ESE allows us to do something that cannot be done in the laboratory. It assumes the nuclei are stationary, so for the moment we consider a very stretched out $H_2$ molecule. If the atoms are distant enough we expect each one to be a normal H atom, for which we know the exact ground state wave function.[1] The singlet wave function for this arrangement might be written

$$^1\psi(1, 2) = N[1s_a(1)1s_b(2) + 1s_b(1)1s_a(2)], \qquad (2.9)$$

where $1s_a$ and $1s_b$ are $1s$ orbital functions centered at nuclei $a$ and $b$, respectively, and $N$ is the normalization constant. This is just the spatial part of the wave function. We may now work with it alone, the only influence left from the spin is the "+" in Eq. (2.9) chosen because we are examining the singlet state. The function of Eq. (2.9) is that given originally by Heitler and London[8].

Perhaps a small digression is in order on the use of the term "centered" in the last paragraph. When we write the ESE and its solutions, we use a single coordinate system, which, of course, has one origin. Then the position of each of the particles, $\vec{r}_i$ for electrons and $\vec{r}_\alpha$ for nuclei, is given by a vector from this common origin. When determining the $1s$ state of H (with an infinitely massive proton), one obtains the result (in au)

$$1s(r) = \frac{1}{\sqrt{\pi}} \exp(-r), \qquad (2.10)$$

where $r$ is the radial distance from the origin of this H atom problem, which is where the proton is. If nucleus $\alpha = a$ is located at $r_a$ then $1s_a(1)$ is a shorthand for

$$1s_a = 1s(|\vec{r}_1 - \vec{r}_a|) = \frac{1}{\sqrt{\pi}} \exp(-|\vec{r}_1 - \vec{r}_a|), \qquad (2.11)$$

and we say that $1s_a(1)$ is "centered at nucleus $a$".

In actuality it will be useful later to generalize the function of Eq. (2.10) by changing its size. We do this by introducing a scale factor in the exponent and write

$$1s'(\alpha, r) = \sqrt{\frac{\alpha^3}{\pi}} \exp(-\alpha r). \qquad (2.12)$$

When we work out integrals for VB functions, we will normally do them in terms of this version of the H-atom function. We may reclaim the real H-atom function any time by setting $\alpha = 1$.

Let us now investigate the normalization constant in Eq. (2.9). Direct substitution yields

$$1 = \langle {}^1\psi(1, 2)| {}^1\psi(1, 2)\rangle \qquad (2.13)$$
$$= |N|^2(\langle 1s_a(1)|1s_a(1)\rangle\langle 1s_b(2)|1s_b(2)\rangle$$
$$+ \langle 1s_b(1)|1s_b(1)\rangle\langle 1s_a(2)|1s_a(2)\rangle$$
$$+ \langle 1s_a(1)|1s_b(1)\rangle\langle 1s_a(2)|1s_b(2)\rangle$$
$$+ \langle 1s_b(1)|1s_a(1)\rangle\langle 1s_b(2)|1s_a(2)\rangle), \qquad (2.14)$$

---

[1] The actual distance required here is quite large. Herring[26] has shown that there are subtle effects due to exchange that modify the wave functions at even quite large distances. In addition, we are ignoring dispersion forces.

where we have written out all of the terms. The $1s$ function in Eq. (2.10) is normalized, so

$$\langle 1s_a(1)|1s_a(1)\rangle = \langle 1s_b(2)|1s_b(2)\rangle, \tag{2.15}$$
$$= \langle 1s_b(1)|1s_b(1)\rangle, \tag{2.16}$$
$$= \langle 1s_a(2)|1s_a(2)\rangle, \tag{2.17}$$
$$= 1. \tag{2.18}$$

The other four integrals are also equal to one another, and this is a function of the distance, $R$, between the two atoms called the overlap integral, $S(R)$. The overlap integral is an elementary integral in the appropriate coordinate system, confocal ellipsoidal–hyperboloidal coordinates[27]. In terms of the function of Eq. (2.12) it has the form

$$S(w) = (1 + w + w^2/3)\exp(-w), \tag{2.19}$$
$$w = \alpha R, \tag{2.20}$$

and we see that the normalization constant for $^1\psi(1, 2)$ is

$$N = [2(1 + S^2)]^{-1/2}. \tag{2.21}$$

We may now substitute $^1\psi(1, 2)$ into the Rayleigh quotient and obtain an estimate of the total energy,

$$E(R) = \langle {}^1\psi(1, 2)|H|{}^1\psi(1, 2)\rangle \geq E_0(R), \tag{2.22}$$

where $E_0$ is the true ground state electronic energy for H₂. This expression involves four new integrals that also can be evaluated in confocal ellipsoidal–hyperboloidal coordinates. In this case all are not so elementary, involving, as they do, expansions in Legendre functions. The final energy expression is ($\alpha = 1$ in all of the integrals)

$$E = 2h - 2\frac{j_1(R) + S(R)k_1(R)}{1 + S(R)^2} + \frac{j_2 + k_2}{1 + S(R)^2} + \frac{1}{R}, \tag{2.23}$$

where

$$h = \alpha^2/2 - \alpha, \tag{2.24}$$
$$j_1 = -[1 - (1 + w)e^{-2w}]/R, \tag{2.25}$$
$$k_1 = -\alpha(1 + w)e^{-w}, \tag{2.26}$$
$$j_2 = [1 - (1 + 11w/8 + 3w^2/4 + w^3/6)e^{-2w}]/R, \tag{2.27}$$
$$k_2 = \{6[S(w)^2(C + \ln w) - S(-w)^2 E_1(4w) + 2S(w)S(-w)E_1(2w)]$$
$$+ (25w/8 - 23w^2/4 - 3w^3 - w^4/3)e^{-2w}\}/(5R), \tag{2.28}$$
$$l_2 = [(5/16 + w/8 + w^2)e^{-w} - (5/16 + w/8)e^{-3w}]/R. \tag{2.29}$$

Of these only the exchange integral of Eq. (2.28) is really troublesome to evaluate. It is written in terms of the overlap integral $S(w)$, the same function of $-w$, $S(-w) = (1 - w + w^2/3)e^w$, the Euler constant, $C = 0.577\,215\,664\,901\,532\,86$, and the exponential integral

$$E_1(x) = \int_x^\infty e^{-y}\frac{dy}{y}, \tag{2.30}$$

which is discussed by Abramowitz and Stegun[28].

In our discussion we have merely given the expressions for the five integrals that appear in the energy. Those interested in the problem of evaluation are referred to Slater[27]. In practice, these expressions are neither very important nor useful. They are essentially restricted to the discussion of this simplest case of the $H_2$ molecule and a few other diatomic systems. The use of AOs written as sums of Gaussian functions has become universal except for single-atom calculations. We, too, will use the Gaussian scheme for most of this book. The present discussion, included for historical reasons, is an exception.

## 2.3 Accuracy of the Heitler–London function

We are now in a position to compare our results with experiment. A graph of $E(R)$ given by Eq. (2.23) is shown as curve (e) in Fig. 2.1. As we see, it is qualitatively correct, showing the expected behavior of having a minimum with the energy rising to infinity at shorter distances and reaching a finite asymptote for large $R$ values. Nevertheless, it misses 34% of the binding energy (comparing with curve (a) of Fig. 2.1), a significant fraction, and its minimum is clearly at too large a bond distance.[2]

## 2.4 Extensions to the simple Heitler–London treatment

In the last section, our calculation used only the function of Eq. (2.9), what is now called the "covalent" bonding function. According to our discussion of linear variation functions, we should see an improvement in the energy if we perform a two-state calculation that also includes the *ionic* function,

$$^1\psi_I(1, 2) = N[1s_a(1)1s_a(2) + 1s_b(1)1s_b(2)]. \tag{2.31}$$

When this is done we obtain the curve labeled (d) in Fig. 2.1, which, we see, represents a small improvement in the energy.

---

[2] The quantity we have calculated here is appropriately compared to $D_e$, the bond energy from the bottom of the curve. This differs from the experimental bond energy, $D_0$, by the amount of energy due to the zero point motion of the vibration. There is no vibration in our system, since the nuclei are infinitely massive. We use the theoretical result for comparison, since it is today considered more accurate than experimental numbers.

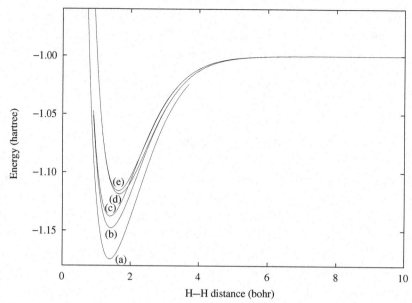

Figure 2.1. Energies of $H_2$ for various calculations using the H-atom $1s$ orbital functions; (b) covalent + ionic, scaled; (c) covalent only, scaled; (d) covalent + ionic, unscaled; (e) covalent only, unscaled. (a) labels the curve for the accurate function due to Kolos and Wolniewicz[31]. This is included for comparison purposes.

A much larger improvement is obtained if we follow the suggestions of Wang[29] and Weinbaum[30] and use the $1s'(1)$ of Eq. (2.12), scaling the $1s$ H function at each internuclear distance to give the minimum energy according to the variation theorem. In Fig. 2.1 the covalent-only energy is labeled (c) and the two-state energy is labeled (b). There is now more difference between the one- and two-state energies and the better binding energy is all but 15% of the total.[3] The scaling factor, $\alpha$, shows a smooth rise from $\approx 1$ at large distances to a value near 1.2 at the energy minimum.

Rhetorically, we might ask, what is it about the ionic function that produces the energy lowering, and just how does it differ from the covalent function? First, we note that the normalization constants for the two functions are the same, and, indeed, they represent exactly the same charge density. Nevertheless, they differ in their two-electron properties. $^1\psi_C$ (adding a subscript $C$ to indicate covalent) gives a higher probability that the electrons be far from one another, while $^1\psi_I$ gives

---

[3] Perhaps we should note that in this relatively simple case, we will approach the binding energy from below as we improve the wave function. This is by no means guaranteed in more complicated systems where both the equilibrium and asymptotic wave functions will be approximate. One of the most important problems of bond energy calculations is to have a "balanced" treatment. This happens if the accuracies at the equilibrium and asymptotic geometries are sufficiently high to give an accurate difference. More realistically in larger systems, one must have any errors at the two geometries sufficiently close to one another to obtain a useful value of the difference. We will take up this question again later.

just the opposite. This is only true, however, if the electrons are close to one or the other of the nuclei. If both of the electrons are near the mid-point of the bond, the two functions have nearly the same value. In fact, the overlap between these two functions is quite close to 1, indicating they are rather similar. At the equilibrium distance the basic orbital overlap from Eq. (2.19) is

$$S = \langle 1s'_s | 1s'_b \rangle = 0.658\,88. \tag{2.32}$$

A simple calculation leads to

$$\Delta = \langle {}^1\psi_C | {}^1\psi_I \rangle = \frac{2S}{1 + S^2} = 0.918\,86. \tag{2.33}$$

(We consider these relations further below.) The covalent function has been characterized by many workers as "overcorrelating" the two electrons in a bond. Presumably, mixing in a bit of the ionic function ameliorates the overage, but this does not really answer the questions at the beginning of this paragraph. We take up these questions more fully in the next section, where we discuss physical reasons for the stability of $H_2$.

At the calculated energy minimum (optimum $\alpha$) the total wave function is found to be

$$\Psi = 0.801\,981\,{}^1\psi_C + 0.211\,702\,{}^1\psi_I. \tag{2.34}$$

The relative values of the coefficients indicate that the variation theorem thinks better of the covalent function, but the other appears fairly high at first glance. If, however, we apply the EGSO process described in Section 1.4.2, we obtain $0.996\,50\,{}^1\psi_C + 0.083\,54\,{}^1\psi'_I$, where, of course, the covalent function is unchanged, but ${}^1\psi'_I$ is the new ionic function orthogonal to ${}^1\psi_C$. On the basis of this calculation we conclude that the the covalent character in the wave function is $(0.996\,50)^2 = 0.993$ (99.3%) of the total wave function, and the ionic character is only 0.7%.

Further insight into this situation can be gained by examining Fig. 2.2, where a geometric representation of the basis vectors and the eigenfunction is given. The overlap integral is the inner product of the two vectors (basis functions) and is the cosine of the angle between them. Since $\arccos(\Delta) = 23.24°$, some care was taken with Fig. 2.2 so that the angle between the vectors representing the covalent and ionic basis functions is close to this value. One conclusion to be drawn is that these two vectors point, to a considerable extent, in the same direction. The two smaller segments labeled (a) and (b) show how the eigenfunction Eq. (2.34) is actually put together from its two components. Now it is seen that the relatively large coefficient of ${}^1\psi_I$ is required because it is poor in "purely ionic" character, rather than because the eigenvector is in a considerably different direction from that of the covalent basis function.

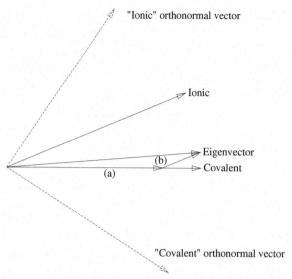

Figure 2.2. A geometric representation of functions for H$_2$ in terms of vectors for $R = R_{eq}$. The small vectors labeled (a) and (b) are, respectively, the covalent and ionic components of the eigenvector. The vectors with dashed lines are the symmetrically orthogonalized basis functions for this case.

It is important to realize that the above geometric representation of the H$_2$ Hilbert space functions is more than formal. The overlap integral of two normalized functions is a real measure of their closeness, as may be seen from

$$\left\langle \left( {}^1\psi_C - {}^1\psi_I \right) \middle| \left( {}^1\psi_C - {}^1\psi_I \right) \right\rangle = 2(1 - \Delta), \tag{2.35}$$

and, if the two functions were exactly the same, $\Delta$ would be 1. As pointed out above, $\Delta$ is a dependent upon $S$, the orbital overlap. Figure 2.3 shows the relation between these two quantities for the possible values of $S$.

In addition, in Fig. 2.2 we have plotted with dashed lines the symmetrically orthogonalized basis functions in this treatment. It is simple to verify that

$$\left\langle {}^1\psi_C - S^1\psi_I \middle| {}^1\psi_I - S^1\psi_C \right\rangle = 0, \tag{2.36}$$

where $S$ is the orbital overlap. Therefore, the vectors given in Fig. 2.2 are just the normalized versions of those in Eq. (2.36). Since they must be at right angles, they must move out 33.38° from the vector they are supposed to approximate. Thus, the real basis functions are closer together than their orthogonalized approximations are to the functions they are to represent. Clearly, writing the eigenfunction in terms of the two symmetrically orthogonalized basis functions will require nearly equal coefficients, a situation giving a very overblown view of the importance of

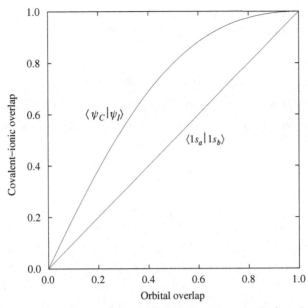

Figure 2.3. The relation between the orbital and covalent–ionic function overlaps.

the "ionic" component of the eigenfunction.[4] Since $\cos 30° = \sqrt{3}/2$, this situation will be the same for any pair of vectors with $S > \sqrt{3}/2$. Only when $S \ll \sqrt{3}/2$, is it likely that the symmetrical orthogonalization will, in the two-vector case, produce coefficients that are qualitatively useful.

Another interesting conclusion can be drawn from Fig. 2.2. The linear combination of AOs-molecular orbital (LCAO-MO) function for H₂ is

$$\Psi_{MO} = 0.510\,46({}^1\psi_C + {}^1\psi_I),\tag{2.37}$$

and, although it has not been shown in the figure to reduce crowding, the vector corresponding to $\Psi_{MO}$ would appear halfway between the two for ${}^1\psi_C$ and ${}^1\psi_I$. It is thus only a little farther from the optimum eigenfunction than ${}^1\psi_C$ alone.

## 2.5 Why is the H₂ molecule stable?

Our discussion so far has not touched upon the origin of the stability of the H₂ molecule. Reading from the articles of the early workers, one obtains the impression that most of them attributed the stability to "resonance" between the $1s_a(1)1s_b(2)$ form of the wave function and the $1s_a(2)1s_b(1)$ form. This phenomenon was new to physicists and chemists at the time and was frequently invoked in "explaining" quantum effects. Today's workers prefer explanations that use classical language,

---

[4] Although the function is more complicated, this is actually what happened in the third column of Table 1.1, where the first two entries are nearly equal compared to those in other columns.

insofar as possible, and attempt to separate the two styles of description to see as clearly as possible just where the quantum effects are.

### 2.5.1 Electrostatic interactions

Equation (2.23) is not well adapted to looking at the nature of the bonding. We rearrange it so that we see the terms that decrease the energy beyond that of two H atoms. This gives

$$E(R) = 2E_H + \frac{J(R) + K(R)}{1 + T(R)}. \tag{2.38}$$

Here $E_H$ is the energy of a normal hydrogen atom, $J(R)$, $K(R)$, and $T(R)$ were called by Heitler and London the Coulomb integral, the exchange integral, and the overlap integral, respectively. The reader should perhaps be cautioned that the terms "Coulomb", "exchange", and "overlap" integrals have been used by many other workers in ways that differ from that initiated by Heitler and London. In the present section we adhere to their original definitions, as follows:

$$J(R) = \langle 1s_a(1)1s_b(2)|V(1, 2)|1s_a(1)1s_b(2)\rangle, \tag{2.39}$$
$$K(R) = \langle 1s_a(1)1s_b(2)|V(1, 2)|1s_b(1)1s_a(2)\rangle, \tag{2.40}$$
$$T(R) = \langle 1s_a(1)1s_b(2)|1s_b(1)1s_a(2)\rangle, \tag{2.41}$$
$$= \langle 1s_a(1)|1s_b(1)\rangle^2,$$

and

$$V(1, 2) = -1/r_{2a} - 1/r_{1b} + 1/r_{12} + 1/R_{ab}. \tag{2.42}$$

These equations are obtained by assigning electron 1 to proton $a$ and 2 to $b$, so that the kinetic energy terms and the Coulomb attraction terms $-1/r_{1a} - 1/r_{2b}$ give rise to the $2E_H$ term in Eq. (2.38). $V(1, 2)$ in Eq. (2.42) is then that part of the Hamiltonian that goes to zero for the atoms at long distances. It is seen to consist of two attraction terms and two repulsion terms. As observed by Heitler and London, the bonding in the $H_2$ molecule arises from the way these terms balance in the $J$ and $K$ integrals. We show a graph of these integrals in Fig. 2.4. Before continuing, we discuss modifications of Eq. (2.38) when scaled $1s$ orbitals are used.

With the $1s'$ function of Eq. (2.12), we obtain

$$E(\alpha, R) = 2E_H + (\alpha - 1)^2 + \alpha \frac{J(\alpha R) + K(\alpha R)}{1 + T(\alpha R)}, \tag{2.43}$$

which reduces to the energy expression of Eq. (2.38) when $\alpha = 1$. The changes brought by including the scale factor are only quantitative in nature and leave the qualitative conclusions unmodified.

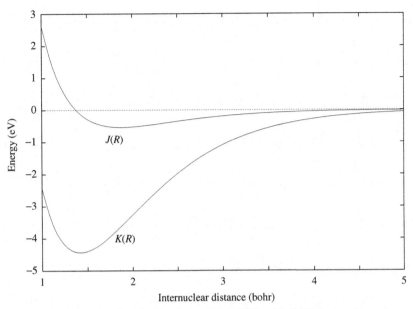

Figure 2.4. The relative sizes of the $J(R)$ and $K(R)$ integrals.

It is important to understand why the $J(R)$ and $K(R)$ integrals have the sizes they do. We consider $J(R)$ first. As we have seen from Eq. (2.42), $V(1, 2)$ is the sum of four different Coulombic terms from the Hamiltonian. If these are substituted into Eq. (2.39), we obtain

$$J(R) = 2j_1(R) + j_2(R) + 1/R,$$
$$j_1(R) = \langle 1s_a| - 1/r_b|1s_a\rangle = \langle 1s_b| - 1/r_a|1s_b\rangle,$$
$$j_2(R) = \langle 1s_a(1)1s_b(2)|1/r_{12}|1s_a(1)1s_b(2)\rangle.$$

The quantity $j_1(R)$ is seen to be the energy of Coulombic attraction between a point charge and a spherical charge distribution, $j_2(R)$ is the energy of Coulombic repulsion between two spherical charge distributions, and $1/R$ is the energy of repulsion between two point charges. $J(R)$ is thus the difference between two attractive and two repulsive terms that cancel to a considerable extent. The magnitude of the charges is 1 in every case. This is shown in Fig. 2.5, where we see that the resulting difference is only a few percent of the magnitudes of the individual terms.

This is to be contrasted with the situation for the exchange integral. In this case we have

$$K(R) = 2k_1(R)S(R) + k_2(R) + S(R)^2/R,$$
$$k_1(R) = \langle 1s_a| - 1/r_b|1s_b\rangle = \langle 1s_a| - 1/r_a|1s_b\rangle,$$
$$k_2(R) = \langle 1s_a(1)1s_b(2)|1/r_{12}|1s_a(2)1s_b(1)\rangle.$$

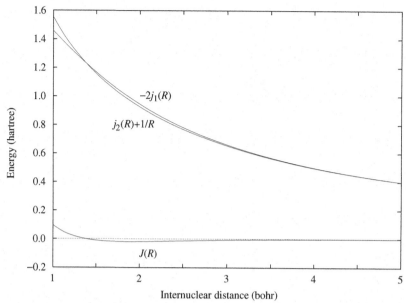

Figure 2.5. Comparison of the sizes of $j_2 + 1/R$ and $-2j_1$ that comprise the positive and negative terms in the Coulomb integral.

The magnitude of the charge in the overlap distribution, $1s_a 1s_b$, is $S(R)$, and here again the overall result is the difference between the energies of attractive and repulsive terms involving the same sized charges of different shaped distributions. The values are shown in Fig. 2.6, where we see that now there is a considerably greater difference between the attractive and repulsive terms. This leads to a value of about 20% of the magnitude of the individual terms.

These values for $J(R)$ and $K(R)$ may be rationalized in purely electrostatic terms involving charge distributions of various sizes and shapes.[5] From the point of view of electrostatics, $J(R)$ is the interaction of points and spherical charge distributions. The well-known effect, where the interaction of a point and spherical charge at a distance $R$ is due only to the portion of the charge inside a sphere of radius $R$, leads to an exponential fall-off $J(R)$, as $R$ increases.

The situation is not so simple with $K(R)$. The overlap charge distribution is shown in Fig. 2.7 and is far from spherical. The upshot of the differences is that the $k_2(R)$ integral is the *self-energy* of the overlap distribution and is more dependent upon its charge than upon its size. In addition, at any distance there is in $k_1(R)$ a portion of the distribution that surrounds the point charge, and, again, the distance dependence is decreased. The overall effect is thus that shown in Fig. 2.4.

---

[5] It should not be thought that the result $|J(R)| \ll |K(R)|$ is peculiar to the $1s$ orbital shape. It is fairly easy to show that a single spherical Gaussian orbital in the place of the $1s$ leads to a qualitatively similar result. In addition, two $sp$ hybrid orbitals, oriented toward one another, show the effect, although compared with spherical orbitals, the disparity between $J$ and $K$ is reduced.

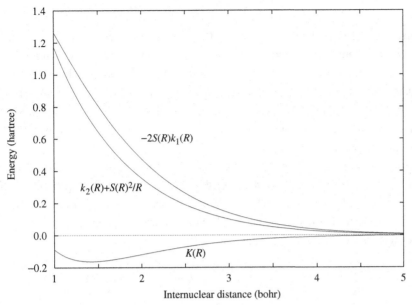

Figure 2.6. Comparison of the sizes of $k_2 + S^2/R$ and $-2k_1S$ that comprise the positive and negative terms in the exchange integral.

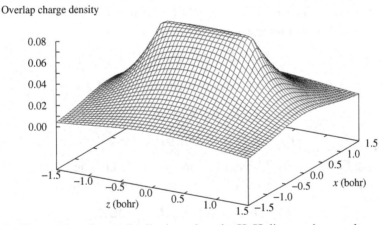

Figure 2.7. The overlap charge distribution when the H–H distance is near the molecular equilibrium value. We show an altitude plot of the value on the $x$–$z$ plane.

We have not yet spoken of the effect of optimizing the scale factor in Eq. (2.43). Wang[29] showed, for the singlet state, that it varies from 1 at $R = \infty$ to about 1.17 at the equilibrium separation. Since both $J$ and $K$ have relatively small slopes near the equilibrium distance, the principal effect is to increase the potential energy portion of the energy by about 17%. The $(\alpha - 1)^2$ term increases by only 3%. Thus the qualitative picture of the bond is not changed by this refinement.

### 2.5.2 Kinetic energy effects

When we go beyond the simple covalent treatment and include an ionic function in the Heitler–London treatment, we obtain a further lowering of the calculated energy. At first glance, perhaps this is surprising, since the ionic function has more electron repulsion than the covalent. Although we saw in Section 1.3.2 that any additional linear variation term must lower the energy, that does not give any physical picture of the process. We will now give a detailed analysis of how the lowering comes about and its physical origin.

In the previous section we examined the variational result of the two-term wave function consisting of the covalent and ionic functions. This produces a $2 \times 2$ Hamiltonian, which may be decomposed into kinetic energy, nuclear attraction, and electron repulsion terms. Each of these operators produces a $2 \times 2$ matrix. Along with the overlap matrix these are

$$S = \begin{bmatrix} 1 & S_{IC} \\ S_{CI} & 1 \end{bmatrix}; \quad T = \begin{bmatrix} T_{II} & T_{IC} \\ T_{CI} & T_{CC} \end{bmatrix};$$

$$V_n = \begin{bmatrix} V_{II} & V_{IC} \\ V_{CI} & V_{CC} \end{bmatrix}; \quad G_e = \begin{bmatrix} G_{II} & G_{IC} \\ G_{CI} & G_{CC} \end{bmatrix}.$$

As we discussed above, the two functions have the same charge density,[6] and this implies that $T_{II} = T_{CC}$ and $V_{II} = V_{CC}$, but we expect $G_{II} > G_{CC}$.

The normalization of the wave function requires

$$1 = C_I^2 + 2S_{IC}C_IC_C + C_C^2. \tag{2.44}$$

Two similar expressions give us the expectation values of $T$ amd $V_n$,

$$\langle T \rangle = C_I^2 T_{II} + 2T_{IC}C_IC_C + C_C^2 T_{CC}, \tag{2.45}$$

$$\langle V_n \rangle = C_I^2 V_{II} + 2V_{IC}C_IC_C + C_C^2 V_{CC}. \tag{2.46}$$

Multiplying Eq. (2.44) by $T_{CC}$ and $V_{CC}$ in turn and subtracting the result from the corresponding Eq. (2.45) or Eq. (2.46), we arrive at the equations

$$\langle T \rangle - T_{CC} = 2(T_{IC} - T_{CC}S_{IC})C_IC_C, \tag{2.47}$$

$$\langle V_n \rangle - V_{CC} = 2(V_{IC} - V_{CC}S_{IC})C_IC_C, \tag{2.48}$$

and we see that the differences depend on how the off-diagonal matrix elements compare to the overlap times the diagonal elements. A similar expression for $G_e$ is more complicated:

$$\langle G_e \rangle - G_{CC} = 2(G_{IC} - G_{CC}S_{IC})C_IC_C + (G_{II} - G_{CC})C_I^2. \tag{2.49}$$

[6] They have the same first order density matrices.

Table 2.1. *Numerical values for overlap, kinetic energy, nuclear attraction,*
*and electron repulsion matrix elements in the two-state calculation.*

|      | $S$       | $T$         | $V$          | $G$         |
|------|-----------|-------------|--------------|-------------|
| II   | 1.0       | 1.146 814 1 | −3.584 134 6 | 0.705 610 0 |
| CI   | 0.933 221 6 | 0.954 081 4 | −3.322 881 7 | 0.600 313 7 |
| CC   | 1.0       | 1.146 814 1 | −3.584 134 6 | 0.584 097 3 |

The numerical values of the matrix elements for $R = R_{min}$ are shown in Table 2.1 Putting in the numbers we see that $T_{IC} - T_{CC}S_{IC} = -0.116\,15$, and, therefore, the kinetic energy decreases as more of the ionic function is mixed in with the covalent. The nuclear attraction term changes in the opposite direction, but by only about one fifth as much, $V_{IC} - V_{CC}S_{IC} = 0.021\,910$. The magnitudes of the numbers in the $G$ column are generally smaller than in the others and we have $G_{IC} - G_{CC}S_{IC} = 0.055\,221$ and $G_{II} - G_{CC} = 0.121\,513$. Since $C_I$ is not very large, the squared term in Eq. (2.49) is not very important. As $C_I$ grows from zero the decrease in the energy is dominated by the kinetic energy until the squared term in Eq. (2.49) can no longer be ignored.

Therefore, the principal role of the inclusion of the ionic term in the wave function is the reduction of the kinetic energy from the value in the purely covalent wave function. Thus, this is the delocalization effect alluded to above. We saw in the last section that the bonding in $H_2$ could be attributed principally to the much larger size of the exchange integral compared to the Coulomb integral. Since the electrical effects are contained in the covalent function, they may be considered a first order effect. The smaller added stabilization due to the delocalization when ionic terms are included is of higher order in VB wave functions.

We have gone into some detail in discussing the Heitler–London treatment of $H_2$, because of our conviction that it is important to understand the details of the various contributions to the energy. Our conclusion is that the bonding in $H_2$ is due primarily to the exchange effect caused by the combination of the Pauli exclusion principle and the required singlet state. The peculiar shape of the overlap distribution causes the exchange effect to dominate. Early texts (see, e.g., Ref. [1]) frequently emphasized the *resonance* between the direct and exchange terms, but this is ultimately due to the singlet state and Pauli principle. Those more familiar with the language of the molecular orbital (MO) picture of bonding may be surprised that the concept of *delocalization energy* does not arise here. That effect would occur in the VB treatment only if ionic terms were included. We thus conclude that delocalization is less important than the exchange attraction in bonding.

## 2.6 Electron correlation

We have pushed this basis to its limit. In fact, it has a basic defect that does not allow a closer approach to the correct answer. The electrons repel each other, and the variation theorem tries to arrange that they not be too close together on the average. This type of effect is called electron correlation, i.e., if the electrons stay away from each other to some extent, their motion is said to be correlated. This calculation does produce some correlation, since we saw that the covalent function tends to keep the electrons apart. This is, however, only in a direction parallel to the bond. When the molecule forms there is also the possibility for angular correlation around the bond direction. Our simple basis makes no provision for this at all, and a significant fraction of the remaining discrepancy is due to this failing. In addition, Rosen[32] added $p_z$ AOs to each center to produce polarization. These, in addition to $p_\pi$ orbitals, will provide more correlation of the type important when the atoms are close as well as correlation of the type generally called van der Waal's forces. We will correct some of these defects in the next section.

## 2.7 Gaussian AO bases

We now turn to considering calculations with the AOs represented by sums of Gaussians. This approach was pioneered by Boys[33], and is used almost universally today. We will settle on a particular basis and investigate its use for a number of different VB-like calculations.

### A double-ζ + polarization basis

We define a ten-function AO basis for the $H_2$ molecule that has two different $s$-type orbitals and one $p$-type set on each H atom. It will be recalled that Weinbaum allowed the scale factor of the $1s$ orbital to adjust at each internuclear distance. Using two "different sized" $s$-type orbitals on each center accomplishes a similar effect by allowing the variation theorem to "choose" the amount of each in the mixture. Our orbitals are shown in Table 2.2. The $s$-type orbitals are a split version of the Huzinaga 6-Gaussian H function and the $p$-type orbitals are adjusted to optimize the energy at the minimum. It will be observed that the $p_\sigma$ and $p_\pi$ scale factors are different. We will present an interpretation of this below.

## 2.8 A full MCVB calculation

The author and his students have used the term multiconfiguration valence bond (MCVB) to describe a linear variation calculation involving more than one VB structure (function). This practice will be continued in the present book. Other terms have been used that mean essentially the same thing[34]. We defer a fuller

Table 2.2. *A double-zeta plus polarization (DZP) basis for $H_2$ calculations.*

| $1s$ | | $1s'$ | | $p_\sigma$ | | $p_\pi$ | |
|---|---|---|---|---|---|---|---|
| exp | $c$ | exp | $c$ | exp | $c$ | exp | $c$ |
| 68.160 0 | 0.002 55 | 0.082 217 | 0.242 60 | 0.9025 | 1.0 | 0.5625 | 1.0 |
| 10.246 5 | 0.019 38 | | | | | | |
| 2.346 48 | 0.092 80 | | | | | | |
| 0.673 320 | 0.294 30 | | | | | | |
| 0.224 660 | 0.492 21 | | | | | | |

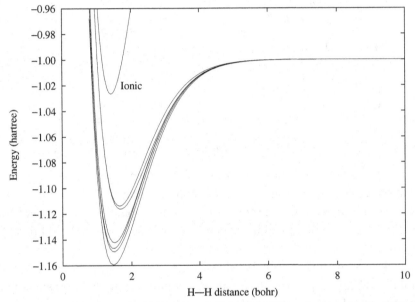

Figure 2.8. A global view of the adiabatic energies of the $H_2$ molecule with the DZP basis.

discussion of these terms until Chapter 7. When the MCVB calculation consists of all of the $n$-electron functions the basis can support, the treatment is said to be *full*.[7]

In the present case we have ten AO basis functions, and these provide a set of 55 symmetric (singlet) spatial functions. Only 27 of these, however, can enter into functions satisfying the spatial symmetry, $^1\Sigma_g^+$, of the ground state of the $H_2$ molecule. Indeed, there are only 14 independent linear combinations for this subspace from the total, and, working in this subspace, the linear variation matrices are only $14 \times 14$. We show the energy for this basis as the lowest energy curve in Fig. 2.8. We will discuss the other curves in this figure later.

---

[7] Compare with the term "full CI" used in the MO regime.

Table 2.3. *Comparison of MCVB coefficients for orthogonalized AOs and "raw" AOs at the equilibrium internuclear distance. The ordering is determined by the orthogonalized AOs.*

| No. | Type | Symmetry function | Orth. AOs | Raw AOs |
|-----|------|-------------------|-----------|---------|
| 1 | C | $(1s_a\ 1s_b)$ | 0.743 681 90 | 0.762 703 95 |
| 2 | I | $(1s_b\ 1s_b)+(1s_a\ 1s_a)$ | 0.144 201 30 | 0.064 562 30 |
| 3 | C | $(1s_a\ 1s_b')+(1s_a'\ 1s_b)$ | −0.104 077 87 | −0.036 499 99 |
| 4 | C | $(1s_a'\ 1s_b')$ | 0.050 698 84 | 0.065 531 68 |
| 5 | I | $(1s_a'\ 1s_a')+(1s_b'\ 1s_b')$ | −0.036 309 53 | −0.069 271 56 |
| 6 | C | $(p_{za}\ 1s_b) - (1s_a\ p_{zb})$ | −0.025 127 35 | −0.034 232 23 |
| 7 | I | $(p_{xb}\ p_{xb})+(p_{yb}\ p_{yb})$ $+(p_{xa}\ p_{xa})+(p_{ya}\ p_{ya})$ | −0.024 702 92 | −0.024 702 92 |
| 8 | I | $(p_{za}\ p_{za})+(p_{zb}\ p_{zb})$ | −0.018 420 72 | −0.018 420 85 |
| 9 | I | $(1s_b'\ p_{zb}) - (1s_a'\ p_{za})$ | 0.016 544 02 | 0.026 116 73 |
| 10 | I | $(1s_a\ 1s_a')+(1s_b\ 1s_b')$ | −0.015 623 66 | 0.137 650 60 |
| 11 | I | $(1s_a\ p_{za}) - (1s_b\ p_{zb})$ | −0.011 973 50 | 0.009 689 91 |
| 12 | C | $(1s_a'\ p_{zb}) - (p_{za}\ 1s_b')$ | −0.011 494 85 | 0.009 130 60 |
| 13 | C | $(p_{xa}\ p_{xb})+(p_{ya}\ p_{yb})$ | −0.007 198 05 | −0.007 198 05 |
| 14 | C | $(p_{za}\ p_{zb})$ | −0.006 660 28 | −0.006 660 48 |

### 2.8.1 Two different AO bases

The Gaussian group functions given in Table 2.2 could be used in "raw" form for our calculation, or we could devise two linear combinations of the raw functions that are orthogonal. The most natural choice for the latter would be a linear combination that is the best H1$s$ orbital and the function orthogonal to it. It should perhaps be emphasized that the energies are identical for the two calculations, except for minor numerical rounding differences. We show the MCVB coefficients for each of these in Table 2.3. The $p$-type orbitals are already orthogonal to the $s$-type and to each other, of course. It will be observed that we orthogonalize only on the same center, not between centers. This is, of course, the *sine qua non* of VB methods.

Examination of the coefficients shows that, although the numbers are not greatly different, there are some significant equalities and differences between the two sets. Considering the equalities first, we note that this occurs for functions 7 and 13. These contribute to angular correlation around the internuclear axis and are completely orthogonal to all of the other functions. This is the reason that the coefficients are the same for the two bases.

At any internuclear separation, the overlap of the raw $s$-type orbitals at the same center is

$$\langle 1s_a | 1s_a' \rangle = 0.709\,09, \tag{2.50}$$

which is fairly large. This produces the greatest difference between the two sets,

Table 2.4. *Comparison of EGSO weights for orthogonalized AOs and "raw" AOs at the equilibrium internuclear distance. The ordering is determined by the orthogonalized AOs.*

| No. | Type | Symmetry function | Orth. AOs | Raw AOs |
|-----|------|-------------------|-----------|---------|
| 1 | C | $(1s_a\ 1s_b)$ | 0.946 195 61 | 0.980 762 89 |
| 10 | I | $(1s_a\ 1s_a') + (1s_b\ 1s_b')$ | 0.043 836 16 | 0.002 080 27 |
| 7 | I | $(p_{xb}\ p_{xb}) + (p_{yb}\ p_{yb})$ $+ (p_{xa}\ p_{xa}) + (p_{ya}\ p_{ya})$ | 0.004 199 01 | 0.004 199 01 |
| 3 | C | $(1s_a\ 1s_b') + (1s_a'\ 1s_b)$ | 0.002 147 20 | [a] |
| 11 | I | $(1s_a\ p_{za}) - (1s_b\ p_{zb})$ | 0.002 071 71 | 0.009 689 91 |
| 2 | I | $(1s_b\ 1s_b) + (1s_a\ 1s_a)$ | 0.000 662 56 | [a] |
| 12 | C | $(1s_a'\ p_{zb}) - (p_{za}\ 1s_b')$ | [a] | 0.003 052 99 |
| 5 | I | $(1s_a'\ 1s_a') + (1s_b'\ 1s_b')$ | [a] | 0.000 654 57 |

[a] These and the functions not listed contribute < 0.01%.

functions 2 and 10. For the orthogonalized set, the $1s_a$ and $1s_b$ are good approximations to the H$1s$ orbital and are important in the ionic function. With the raw AOs the most important ionic term becomes function 10, which mixes the two types. The fact that the function 1 coefficient is larger for the raw AOs than for the other basis should not be considered too important. There is a rather larger amount of overlap[8] for the former basis.

We now compare the EGSO weights of the important functions between the orthogonalized and raw bases. These are shown in Table 2.4. At first glance, in looking at the covalent function, one might be surprised at how much larger the weight is for the raw orbital. A little reflection will show, however, that this is to be expected and is related to the way these calculations accomplish the effects of orbital scaling. It will be recalled that the orbital scale factor in Eq. (2.12) was optimally $\approx 1.2$ at the equilibrium internuclear separation in the simple calculation. The raw AO called $1s$ in Table 2.4 does not have the long range component contained in the optimized AO. Therefore, the raw AO is "tighter" and the closer of the two to the AO the molecule desires at $R_{eq}$. On the other hand the orthogonalized $1s$ AO is the function appropriate for long range. The "ionic" function $(1s_a\ 1s_a') + (1s_b\ 1s_b')$, which contains both short and long range orthogonalized AOs, compensates for the too diffuse character of the orthogonalized $1s$. We note that the sum of the first two weights is 0.990 031 77 for orthogonalized AOs and 0.982 843 16 for raw AOs. These are not too far from one another and indicate a similar representation of the total wave function.

---

[8] This is measured by the determinant of $14 \times 14$ MCVB overlap matrix. The smaller this is, the larger are all of the coefficients of the VB functions.

Table 2.5. *Comparison of MCVB coefficients for orthogonalized*
*AOs and raw AOs at the internuclear distance of 20 bohr.*

| No. | Type | Symmetry function | Orth. AOs | Raw AOs |
|-----|------|-------------------|-----------|---------|
| 1 | C | $(1s_a \, 1s_b)$ | 1.000 000 00 | 0.661 346 76 |
| 2 | C | $(1s_a \, 1s_b') + (1s_a' \, 1s_b)$ | 0.000 000 00 | 0.197 293 67 |
| 3 | C | $(1s_a' \, 1s_b')$ | 0.000 000 00 | 0.058 856 86 |

Table 2.6. *Comparison of EGSO weights for orthogonalized*
*and raw AOs at an internuclear distance of 20 bohr.*

| No. | Type | Symmetry function | Orth. AOs | Raw AOs |
|-----|------|-------------------|-----------|---------|
| 1 | C | $(1s_a \, 1s_b)$ | 1.000 000 00 | 0.942 329 44 |
| 2 | C | $(1s_a \, 1s_b') + (1s_a' \, 1s_b)$ | 0.000 000 00 | 0.056 814 21 |
| 3 | C | $(1s_a' \, 1s_b')$ | 0.000 000 00 | 0.000 856 35 |

When we make these same comparisons for an internuclear separation of 20 bohr, we obtain the coefficients shown in Table 2.5 and the weights shown in Table 2.6. Now the orthogonalized AOs give the asymptotic function with one configuration, while it requires three for the raw AOs. The energies are the same, of course. The EGSO weights imply the same situation. A little reflection will show that the three terms in the raw VB function are just those required to reconstruct the proper H$1s$ orbital.

It should be clear that coefficients and weights in such calculations as these depend on the exact arrangement of the basis, and that their interpretations also depend upon how much physical or chemical significance can be associated with individual basis functions.

### 2.8.2 Effect of eliminating various structures

As we stated above, there are 14 different symmetry functions in the full MCVB with the present basis we are discussing. It will be instructive to see how the adiabatic energy curve changes as we eliminate these various functions in a fairly systematic way. This is the source of the higher-energy curves in Fig. 2.8. We showed all of them there in spite of the fact that they are not all easily distinguishable on that scale, because that gives a better global view of how they change. We "blow up" the region around the minimum and show this in Fig. 2.9 where the six lowest ones are labeled (a)–(f). In addition, the Kolos and Wolniewicz curve is shown for comparison and marked "K&W".

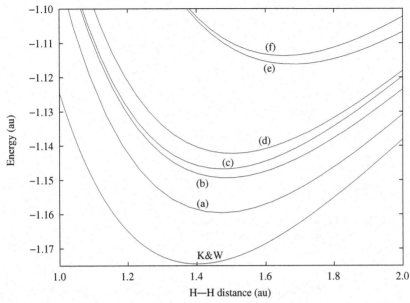

Figure 2.9. A detailed view of the adiabatic energies of the $H_2$ molecule with the DZP basis.

(a) This is the full calculation with all of the functions of Table 2.3.
(b) This has had functions 7 and 13 (see Table 2.3) removed. These give angular correlation around the internuclear axis.
(c) This has had functions 4 and 5 eliminated (in addition to 7 and 13). This clearly does not have much effect. We can categorize different structures according to how many electrons are in orbitals that might be considered "excited". In this case all of the AOs except the two $H1s$ orbitals will be considered excited. Functions 7 and 13 are of a sort we will call "double excitations" and do not contribute much to the energy.
(d) Now 6, 8, 9, 11, 12, and 14 have been eliminated. These are the configurations that contain the $p_\sigma$ AO. These are fairly important, it is seen.
(e) When we now eliminate 3 and 10 and we are left with only the covalent and ionic configurations of the asymptotic $H1s$ function.
(f) This is only function 1, the covalent one. It is seen that the ionic term contributes fairly little at this level.

The energy increments from omitting selected basis functions are not additive, thus, using the amount that the energy is raised by elimination as a measure of the importance of various configurations is not a unique process, since the result depends upon the order of elimination. Nevertheless, the previous exercise was instructive.

Although one would not expect a good answer, we show one more calculation with a limited basis – that is the $H1s$ "ionic" function by itself. It does not fall in the region plotted in Fig. 2.9, but is shown in Fig. 2.8 where it is marked "Ionic".

Table 2.7. *A TZ2P1D$^a$ basis for H$_2$ calculations.*

| s, s', s'' | | p, p' | | d | |
|---|---|---|---|---|---|
| exp$^b$ | c$^c$ | exp$^b$ | c$^c$ | exp$^b$ | c$^c$ |
| 68.160 0 | 0.002 55 | 5.027 243 | 0.092 05 | 1.140 046 | 1.0 |
| 10.246 5 | 0.019 38 | 1.190 621 | 0.474 06 | | |
| 2.346 48 | 0.092 80 | | | | |
| 0.673 320 | 0.294 30 | 0.450 098 | 0.578 60 | | |
| 0.224 660 | 0.492 21 | | | | |
| 0.082 217 | 0.242 60 | | | | |

$^a$ Triple-zeta + 2 $p$-functions + 1 $d$-function.
$^b$ exp = exponential scale factor.
$^c$ $c$ = coefficient.

### 2.8.3 Accuracy of full MCVB calculation with 10 AOs

The full MCVB calculation gives the best answer we have obtained so far. Compared to the Kolos and Wolniewicz result we now have 91.5% of the binding energy, but the minimum is at 0.778 instead of 0.741 Å, almost 5% too large. One must realize that the difficulty here is *not* with the VB method, but, rather, with the underlying AO basis. We are evaluating the energy for the full calculation, which would be the same whether we are using the VB method, orthogonal MOs followed by a full configuration interaction (CI), or some combination.[9]

### 2.8.4 Accuracy of full MCVB calculation with 28 AOs

It is instructive to increase the size of the AO basis to see where we get to in calculating the binding energy of H$_2$. This is a so-called triple-$\zeta$ basis with a split $p$ set and a $d$ set on each center. It is shown in Table 2.7 and is based upon the same six-function Huzinaga orbital as is used in the previous Gaussian basis, Table 2.2. There are 406 singlet functions that can be made from this basis, but only 128 of them can enter into $^1\Sigma_g^+$ molecular states, and these give 58 linearly independent $^1\Sigma_g^+$ functions. The $1s$, $2s$, $3s$, $1p$, and $2p$ orbitals we use are the eigenfunctions of the H-atom Hamiltonian matrix in the $s$, $s'$, $s''$, $p$, and $p'$ group function basis. There is only one $d$-function, and it needs no modification.

The results are considerably improved over the basis of Table 2.7. We now obtain 98.6% of the binding energy and the minimum is at 0.7437 Å, which is only 0.3%

---

[9] It is perhaps not too difficult to see that a nonsingular linear transformation of the underlying AO basis produces a nonsingular linear transformation of the $n$-electron basis. Thus, the $H$ and $S$ matrices imply the same eigenvalues, although the coefficients in the sum giving the wave function will differ. Nevertheless, the actual wave function for a given eigenvalue (nondegenerate ones, at least) will be the same.

Table 2.8. *EGSO weights (>0.001) for two bases.*

| | 10 AOs | | | 28 AOs |
|---|---|---|---|---|
| 1 | $(1s_a\ 1s_b)$ | 0.946 195 61(C) | 0.947 336 55(C) | $(1s_a\ 1s_b)$ |
| 2 | $(1s_a\ 2s_a)+(1s_b\ 2s_b)$ | 0.043 836 16(I) | 0.030 228 93(I) | $(1s_a\ 2s_a)+(1s_b\ 2s_b)$ |
| 3 | | | 0.012 119 46(C) | $(1s_a\ 3s_b)+(1s_b\ 3s_a)$ |
| 4 | $(p_{xa}\ p_{xa})+(p_{ya}\ p_{ya})$ $+(p_{xb}\ p_{xb})+(p_{yb}\ p_{yb})$ | 0.004 199 01(I) | 0.003 736 22(I) | $(1p_{xa}\ 1p_{xa})+(1p_{ya}\ 1p_{ya})$ $+(1p_{xb}\ 1p_{xb})+(1p_{yb}\ 1p_{yb})$ |
| 5 | $(1s_a\ 2s_b)+(2s_a\ 1s_b)$ | 0.002 147 20(C) | | |
| 6 | | | 0.002 316 93(C) | $(1s_a\ 1p_{zb})+(1s_b\ 1p_{za})$ |
| 7 | $(1s_a\ p_{za})+(1s_b\ p_{zb})$ | 0.002 071 71(I) | | |
| | Total | 0.998 449 70 | 0.995 738 09 | Total |

too large. We could improve these results further, but for our purposes in discussing VB theory this is not particularly pertinent. Rather, we compare the EGSO weights of the two calculations to ascertain how much they change.

### 2.8.5 EGSO weights for 10 and 28 AO orthogonalized bases

In Table 2.8 we show a comparison of the EGSO weights for the two full MCVB calculations we have made with orthogonalized Gaussian bases. These are quite close to one another. We have only listed functions with weights > 0.001, and in each case there are five.

We can interpret the various weights as follows.

1. **Covalent** The principal function in each case is the conventional Heitler–London covalent basis function with a weight very close to 95%.
2. **Ionic** The function, in each case, with the next highest weight, 3–4%, is ionic and involves a single excitation into the $2s$ AO. This contributes to adjusting the electron correlation and also contributes to adjusting the size of the wave function along the lines of the scale adjustment of the Weinbaum treatment. As we have shown, it also contributes to delocalization.
3. **Covalent** This function at 1.2% appears only with the larger basis set involving, as it does, the higher $3s$-function. It will contribute to scaling.
4. **Ionic** At $\approx$0.4% the next function type appears in both sets and contributes to the angular correlation around the internuclear line.
5. **Covalent** At $\approx$0.2% the next function appears only with the smaller basis. It is the counterpart of the $3s$ covalent function with the larger basis, but is relatively less important.
6. **Covalent** At $\approx$0.2% this function contributes to polarization with the larger basis.
7. **Ionic** Again, at $\approx$0.2% this function contributes to polarization with the smaller basis.

These functions contribute over 99% of the total wave function in both cases. The smaller value for the larger basis reflects the larger number of small contribution basis functions in that case. Although the fairly large number of basis functions that contribute with only minor weights have an important impact on lowering the energy, the large weight of the covalent function indicates that the bond in the $H_2$ molecule is just as chemists always describe it: a covalent bond.

# 3

# H$_2$ and delocalized orbitals

We now examine VB functions where the orbitals are allowed to take much more general forms, but only one configuration is used. This more general form allows the orbitals to range over more than one atomic center. As we shall see later, the restriction to one configuration is appropriate only to two-electron systems, so we must postpone the discussions of more configurations until we treat the more advanced methods in Chapter 5.

## 3.1 Orthogonalized AOs

Before we examine the more general case, let us look at an unusual result due to Slater. Earlier, in discussing solids Wannier[35] had shown how linear combinations of the AOs could be made that rendered the functions orthogonal while retaining a relatively large concentration on one center. Slater adapted this idea to the H$_2$ molecule. In modern language this is just making a symmetric orthogonalization (see Section 1.4.2) of the basis, which in this case is a H1$s$ function on each of two centers, $1s_a$ and $1s_b$. We are here again, following Slater, using the correct exponential functions of Eq. (2.10). The overlap matrix for this basis is

$$\bar{S} = \begin{bmatrix} 1 & S \\ S & 1 \end{bmatrix},$$  (3.1)

and the inverse square root is

$$\bar{S}^{-1/2} = \begin{bmatrix} \dfrac{1}{2\sqrt{1+S}} + \dfrac{1}{2\sqrt{1-S}} & \dfrac{1}{2\sqrt{1+S}} - \dfrac{1}{2\sqrt{1-S}} \\ \dfrac{1}{2\sqrt{1+S}} - \dfrac{1}{2\sqrt{1-S}} & \dfrac{1}{2\sqrt{1+S}} + \dfrac{1}{2\sqrt{1-S}} \end{bmatrix},$$  (3.2)

where $S = \langle 1s_a | 1s_b \rangle$, and the signs are appropriate for $S > 0$. This orthogonalization gives us two new functions (see Eq. (1.48))

$$|A\rangle = P|1s_a\rangle + Q|1s_b\rangle, \tag{3.3}$$
$$|B\rangle = Q|1s_a\rangle + P|1s_b\rangle, \tag{3.4}$$

where

$$P = \frac{1}{2\sqrt{1+S}} + \frac{1}{2\sqrt{1-S}}, \tag{3.5}$$

$$Q = \frac{1}{2\sqrt{1+S}} - \frac{1}{2\sqrt{1-S}}. \tag{3.6}$$

We use these in a single Heitler–London covalent configuration,

$$\Psi_{orth} = A(1)B(2) + B(1)A(2),$$

and calculate the energy. When $R \to \infty$ we obtain $E = -1$ au, just as we should. At $R = 0.741$ Å, however, where we have seen that the energy should be a minimum, we obtain $E = -0.6091$ hartree, much higher than the correct value of $-1.1744$ hartree. The result for this orthogonalized basis, which represents not only no binding but actual repulsion, could hardly be worse.

It is interesting to consider this function in terms of the covalent and ionic functions of Chapter 2. If the $|A\rangle$ and $|B\rangle$ functions in terms of the basic AOs are substituted into $\Psi_{orth}$ and the result normalized, one sees that

$$\Psi_{orth} = \frac{\sqrt{1+S^2}}{1-S^2}(\psi_C - S\psi_I),$$

where, as always, $S$ is the orbital overlap. Thus, this is exactly the symmetrically orthogonalized function closer to $\psi_C$ discussed above, and its vector representation in Fig. 2.2 is clearly a considerable distance from the optimum eigenvector. Thus we should not be surprised at the poor value for the variational energy corresponding to $\Psi_{orth}$.

The early workers do not comment particularly on this result, but, in light of present understanding, we may say that the symmetric orthogonalization gives very close to the poorest possible linear combination for determining the lowest energy. This results from the added kinetic energy of the orbitals produced by a node that is not needed. Alternatively, one could say that the symmetric orthogonalization yields *antibonding* orbitals where bonding orbitals are needed. This is a good example of how the orthogonalization between different centers can have serious consequences for obtaining good energies and wave functions. We shall see shortly that there are, however, linear combinations determined in other ways that work quite well.

## 3.2 Optimal delocalized orbitals

We now investigate orbitals that range over both centers with linear combinations that minimize the calculated energy. For this simple two-electron system these may all be viewed as extensions of the Coulson–Fisher approach we describe next. We use the basis of Table 2.2 and compare these results with the appropriate full MCVB calculations of Section 2.8.

### 3.2.1 The method of Coulson and Fisher[15]

The first calculation of the energy of $H_2$ for optimal delocalized orbitals used

$$A = 1s_a + \lambda 1s_b, \tag{3.7}$$

$$B = \lambda 1s_a + 1s_b, \tag{3.8}$$

and, using the "covalent" function, $A(1)B(2) + B(1)A(2)$, in the Rayleigh quotient, adjusted the value of $\lambda$ to minimize the energy. We will not duplicate this calculation here, but bring this up, because the methods we will discuss are generalizations of the Coulson–Fisher approach where we use in the orbitals all of the functions of our basis with the appropriate symmetry.

### 3.2.2 Complementary orbitals

It will be observed that the Coulson–Fisher functions satisfy the relations

$$\sigma_h A = B \tag{3.9}$$

and

$$\sigma_h B = A, \tag{3.10}$$

where $\sigma_h$ is the operation of $D_{\infty h}$ that reflects the molecule end for end. If $A$ and $B$ are also of $\sigma$ symmetry,[1] the "covalent" function $A(1)B(2) + B(1)A(2)$ is of $^1\Sigma_g^+$ symmetry. Thus, the overall state symmetry is correct, although the orbitals do not belong to a single irreducible representation. For our first calculation we take all of the $\sigma$-type AOs of the basis and form (the unnormalized)

$$A = 1s_a + a1s_b + b2s_a + c2s_b + dp_{za} + ep_{zb} \tag{3.11}$$

and

$$B = \sigma_h A, \tag{3.12}$$

---

[1] The reader should note carefully the two different uses of the symbol $\sigma$ here. One is a group operation, the other the state designation of an orbital.

Table 3.1. *Energies of optimal orbital calculations.*

| Distance Å | Complementary orbitals | Unsymmetric orbitals | Full MCVB |
|---|---|---|---|
| 0.741 | −1.143 356 | −1.147 368 | −1.148 052 |
| ∞ | −1.0 | −1.0 | −1.0 |

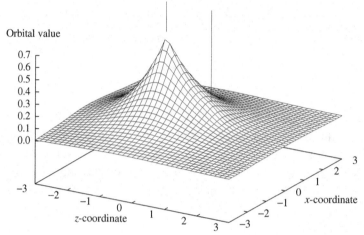

Figure 3.1. Altitude drawing of the $A$ optimal complementary orbital for values in the $x$–$z$ plane. The H nuclei are on the $z$-axis. The two vertical lines point at the nuclei.

where $a, \ldots, e$ are the variation constants to be optimized.[2] These orbitals are inserted in the covalent function, and the Rayleigh quotient minimized with respect to the variation parameters. We show the results for two internuclear distances in the second and fourth columns of Table 3.1 together with the calculation of the full MCVB using the same AO basis, i.e., omitting the $p_\pi$ AOs.

It will be recalled by examining Table 2.3 that there are 12 independent $\sigma$-AO-only VB functions in the MCVB. Our complementary orbital function has only five independent parameters, so it certainly cannot duplicate the MCVB energy, but it reproduces 96.8% of the binding energy of the latter calculation.

We show a 3D altitude drawing of the amplitude of the $A$ orbital in Fig. 3.1. It is easily seen to be extended over both nuclei, and it is this property that produces in the wave function the adjustment of the correlation and delocalization that is provided by the ionic function in the linear variation treatment with the same AO basis.

We point out that these results are obtained without any "ionic" states in the wave function and such are not needed. As we argued in Chapter 2, the principal role of the ionic functions is to provide delocalization of the electrons when the molecule

---

[2] We note that we cannot introduce the $p_\pi$ AOs here and retain the $^1\Sigma_g^+$ state symmetry.

forms. Since the orbital is itself delocalized, the wave function requires nothing further.

We also anticipate the discussion of Chapter 7 by pointing out that the wave function we have obtained here is a simple version of Goddard's generalized valence bond (GGVB) or the spin coupled valence bond SCVB treatment of Gerratt, Cooper, and Raimondi. The GGVB in general has orthogonality prescriptions that do not, however, arise in the two electron case.

### 3.2.3 Unsymmetric orbitals

Instead of using Eq. (3.12) we might use a $B$ defined as

$$B' = 1s_b + a' \, 1s_a + b' \, 2s_b + c' \, 2s_a + d' \, p_{zb} + e' \, p_{za}. \qquad (3.13)$$

Of course, if this is used in $A(1)B'(2) + B'(1)A(2)$, the result does not have the correct symmetry, therefore we must use a projection operator to obtain the $^1\Sigma_g^+$ state. Defining $A' = \sigma_h B'$ and $B = \sigma_h A$, we have

$$\frac{1}{2}[I + \sigma_h][A(1)B'(2) + B'(1)A(2)]$$

$$= \frac{1}{2}[A(1)B'(2) + B'(1)A(2) + B(1)A'(2) + A'(1)B(2)], \qquad (3.14)$$

and when this ten-parameter function is optimized with the Rayleigh quotient we obtain the results in the third column of Table 3.1 We now have 99.5% of the full binding energy, which is a credible showing. These orbitals are visibly different from the complementary optimal orbital as can be seen in the plots of $A$ in Fig. 3.2

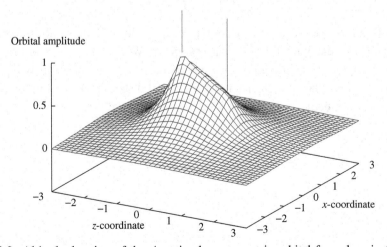

Figure 3.2. Altitude drawing of the $A$ optimal unsymmetric orbital for values in the $x$–$z$ plane. The H nuclei are on the $z$-axis. The two vertical lines point at the nuclei.

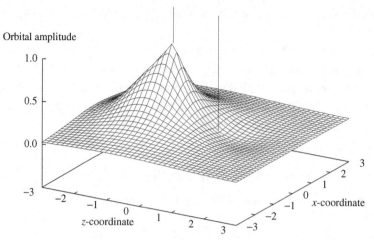

Figure 3.3. Altitude drawing of the $A'$ optimal unsymmetric orbital for values in the $x-z$ plane. The H nuclei are on the $z$-axis. The two vertical lines point at the nuclei.

and $A'$ in Fig. 3.3. This $A$ is more evenly distributed over the molecule, and $A'$ is less so, being somewhat outside of the region between the nuclei.

The question might be asked: can angular correlation be included in an optimal orbital treatment? The answer is yes, but it is somewhat troublesome in general with infinite groups like $D_{\infty h}$. We merely need to generalize the trick we pulled to obtain the wave function of Eq. (3.14). The projection operator there guarantees the $g$ (gerade) subscript on the state symmetry, $^1\Sigma_g^+$. If we add $p_\pi$ orbitals to our unsymmetric optimal orbitals we must also apply a projection operator to guarantee the $\Sigma$ part of the state symmetry. The appropriate operator in general is then

$$\frac{1}{4\pi}(I + \sigma_h) \int_0^{2\pi} C_\phi d\phi, \tag{3.15}$$

where $C_\phi$ is a rotation about the $z$-axis of $\phi$ radians. This is not an operation convenient to deal with on a digital computer. We will not pursue these ideas further. As stated, such integral projection operators are troublesome to implement, and, in particular, they are clearly not very useful if there is no symmetry, which is true of most molecules.

# 4

# Three electrons in doublet states

In Chapter 5 we give an analysis of VB functions that is general for any number of electrons. In order to motivate some of the considerations we discuss there we first give a detailed example of the requirements when one is to construct an antisymmetric doublet eigenfunction of the spin for a three-electron system. Pauncz[36] has written a useful workbook on this subject.

We will first give a discussion of some results of general spin-operator algebra; not much is needed. This is followed by a derivation of the requirements spatial functions must satisfy. These are required even of the exact solution of the ESE. We then discuss how the orbital approximation influences the wave functions. A short qualitative discussion of the effects of dynamics upon the functions is also given.

## 4.1 Spin eigenfunctions

The total spin operator and operator for the $z$-component are

$$\vec{S}^2 = \vec{S}_1^2 + \vec{S}_2^2 + \vec{S}_3^2, \tag{4.1}$$

$$S_z = S_{z1} + S_{z2} + S_{z3}, \tag{4.2}$$

where we see that both operators are *symmetric*[1] sums of operators for the three identical electrons. Many treatments of spin discuss the raising and lowering operators for the $z$-component of the total spin[4]. These are symmetric operators we symbolize as

$$S^+ = S_x + iS_y \tag{4.3}$$

for raising and

$$S^- = S_x - iS_y \tag{4.4}$$

---

[1] The term *symmetric* is used in a variety of ways by mathematicians and in this book. The important point here is that the term implies that for $n$ particles these spin operators commute with any permutation of $n$ objects.

for lowering. With these, two alternative forms of $\vec{S}^2$ are possible:

$$\vec{S}^2 = S^-S^+ + S_z(S_z + 1), \tag{4.5}$$
$$= S^+S^- + S_z(S_z - 1). \tag{4.6}$$

These are quite useful for constructing spin eigenfunctions and are easily seen to be true, not only for three electrons, but for $n$.

In Chapter 2 we used $\eta_{\pm 1/2}$ to represent individual electron spin functions, but we would now like to use a more efficient notation. Thus we take $[+ + +]$ to represent the product of three $m_s = +1/2$ spin functions, one for electron 1, one for electron 2, and one for electron 3. As part of the significance of the symbol we stipulate that the $+$ or $-$ signs refer to electrons 1, 2, and 3 in that order. Thus, in the notation of Chapter 2, we have, for example,

$$[+ + -] = \eta_{1/2}(1)\eta_{1/2}(2)\eta_{-1/2}(3). \tag{4.7}$$

Familiar considerations show that there are all together eight different $[\pm \pm \pm]$, they are all normalized and mutually orthogonal, and they form a complete basis for spin functions of three electrons.

The significance of Eqs. (4.5) and (4.6) is that *any* spin function $\phi$ with the properties

$$S^+\phi = 0 \tag{4.8}$$

and

$$S_z\phi = M_S\phi, \tag{4.9}$$

is automatically also an eigenfunction of the total spin with eigenvalue $M_S(M_S + 1)$. Similarly, if $\phi$ satisfies

$$S^-\phi = 0 \tag{4.10}$$

and

$$S_z\phi = M_S\phi, \tag{4.11}$$

it is automatically also an eigenfunction of the total spin with eigenvalue $M_S$ $(M_S - 1)$.

We may use this to construct doublet eigenfunctions of the total spin for our three electrons. Thus, consider

$$\phi = a[- + +] + b[+ - +] + c[+ + -], \tag{4.12}$$

where, clearly, we have $S_z\phi = \frac{1}{2}\phi$. Applying the operator $S^+$ to this gives

$$S^+\phi = (a + b + c)[+ + +]. \tag{4.13}$$

According to our requirements, this must be zero if $\phi$ is to be an eigenfunction of the total spin, therefore, we must have $(a + b + c) = 0$, since $[+ + +]$ certainly is not zero. We have one (homogeneous) equation in three unknowns so there is more than one solution – in fact, there are an infinite number of solutions. Nevertheless, all of them may be written as linear combinations of (in this case) just two. We observe that we can write three solutions of the form $(a, b, c) = (1, -1, 0), (1, 0, -1)$, and $(0, 1, -1)$, but that any one of these may be written as the difference of the other two. Thus, there are only two *linearly independent* solutions among our three, and *any* doublet spin function for three electrons may be written as a linear combination of these two.

When dealing with spin functions it is normally convenient to arrange the bases to be orthonormal, and we obtain two functions,

$$^2\phi_1 = \frac{1}{\sqrt{6}}(2[+ + -] - [+ - +] - [- + +]) \tag{4.14}$$

and

$$^2\phi_2 = \frac{1}{\sqrt{2}}([+ - +] - [- + +]). \tag{4.15}$$

For simplicity we do not label these functions with the $M_S$ value. Our work in VB theory and solving the ESE seldom needs any but the principal spin function with $M_S = S$. The $S^-$ operator is always available should other $M_S$ values be needed.

With the spin eigenfunctions of Eqs. (4.14) and (4.15) we have an example of the *spin degeneracy* alluded to in Chapter 2. Unlike the single singlet function we arrived at for two electrons in Section 2.1.1 we now obtain two.[2] Writing out the equations specifically,

$$\vec{S}^2 \, ^2\phi_1 = \tfrac{1}{2}(\tfrac{1}{2} + 1)^2 \phi_1, \tag{4.16}$$

$$\vec{S}^2 \, ^2\phi_2 = \tfrac{1}{2}(\tfrac{1}{2} + 1)^2 \phi_2, \tag{4.17}$$

we see that both of the functions have the same eigenvalue, and it is degenerate. In Chapter 5 we shall see that the degree of this degeneracy is related to the sizes of irreducible representations of the symmetric groups. We defer further discussion until that place.

## 4.2 Requirements of spatial functions

We now have a significant difference from the case of two electrons in a singlet state, namely, we have two spin functions to combine with spatial functions for a solution to the ESE rather than only one. For a doublet three-electron system our general solution must be

$$^2\Psi = \, ^2\psi_1 \, ^2\phi_1 + \, ^2\psi_2 \, ^2\phi_2, \tag{4.18}$$

---

[2] Four really, considering that each $^2\phi_1$ and $^2\phi_2$ has both $m_S = \pm\tfrac{1}{2}$ forms, also.

and, apparently, we have two spatial functions to determine. We have not yet applied the antisymmetry requirement, however. With this it will develop that $^2\psi_1$ and $^2\psi_2$ are not really independent and only one need be determined.

We must now investigate the effect of the binary interchange operators, $P_{ij}$ on the $^2\phi_i$ functions. We suppress the $^2$ spin-label superscript for these considerations. It is straightforward to determine

$$P_{12}\phi_1 = \phi_1, \tag{4.19}$$

$$P_{12}\phi_2 = -\phi_2, \tag{4.20}$$

$$P_{13}\phi_1 = (2[-++] - [+-+] - [++-])/\sqrt{6}$$

$$= -\frac{1}{2}\phi_1 - \frac{\sqrt{3}}{2}\phi_2, \tag{4.21}$$

$$P_{13}\phi_2 = ([+-+] - [++-])/\sqrt{2}$$

$$= -\frac{\sqrt{3}}{2}\phi_1 + \frac{1}{2}\phi_2, \tag{4.22}$$

$$P_{23}\phi_1 = (2[+-+] - [++-] - [-++])/\sqrt{6}$$

$$= -\frac{1}{2}\phi_1 + \frac{\sqrt{3}}{2}\phi_2, \tag{4.23}$$

$$P_{23}\phi_2 = ([++-] - [-++])/\sqrt{2}$$

$$= \frac{\sqrt{3}}{2}\phi_1 + \frac{1}{2}\phi_2, \tag{4.24}$$

and the results of applying higher permutations may be determined from these.

We now apply the $P_{ij}$ operators to $\Psi$ and require the results to be antisymmetric. Using the fact that the $\phi_i$ are linearly independent, for $P_{12}$ we obtain

$$P_{12}\Psi = -\Psi = (P_{12}\psi_1)\phi_1 - (P_{12}\psi_2)\phi_2,$$

$$P_{12}\psi_1 = -\psi_1, \tag{4.25}$$

$$P_{12}\psi_2 = \psi_2, \tag{4.26}$$

and the others in a similar way give

$$P_{13}\psi_1 = \frac{1}{2}\psi_1 + \frac{\sqrt{3}}{2}\psi_2, \tag{4.27}$$

$$P_{13}\psi_2 = \frac{\sqrt{3}}{2}\psi_1 - \frac{1}{2}\psi_2, \tag{4.28}$$

$$P_{23}\psi_1 = \frac{1}{2}\psi_1 - \frac{\sqrt{3}}{2}\psi_2, \tag{4.29}$$

$$P_{23}\psi_2 = -\frac{\sqrt{3}}{2}\psi_1 - \frac{1}{2}\psi_2, \tag{4.30}$$

and, finally,

$$P_{12}P_{13}\psi_1 = -\frac{1}{2}\psi_1 + \frac{\sqrt{3}}{2}\psi_2, \tag{4.31}$$

$$P_{12}P_{13}\psi_2 = -\frac{\sqrt{3}}{2}\psi_1 - \frac{1}{2}\psi_2, \tag{4.32}$$

$$P_{12}P_{23}\psi_1 = -\frac{1}{2}\psi_1 - \frac{\sqrt{3}}{2}\psi_2, \tag{4.33}$$

$$P_{12}P_{23}\psi_2 = \frac{\sqrt{3}}{2}\psi_1 - \frac{1}{2}\psi_2. \tag{4.34}$$

With all of these relations it is not surprising that we can find several that express $\psi_2$ in terms of sums of permuted $\psi_1$ functions. An example is

$$\psi_2 = (P_{13} - P_{23})\psi_1/\sqrt{3}. \tag{4.35}$$

This allows us to obtain some information about the normalization of the $\psi_i$ functions, since

$$\langle\psi_2|\psi_2\rangle = \frac{1}{3}\langle(P_{13} - P_{23})\psi_1|(P_{13} - P_{23})\psi_1\rangle,$$

$$= \frac{1}{3}\langle\psi_1|(I - P_{13}P_{23} - P_{23}P_{13} + I)\psi_1\rangle,$$

$$= \frac{1}{3}\langle\psi_1|(2I - P_{12}P_{13} - P_{12}P_{23})\psi_1\rangle,$$

$$= \langle\psi_1|\psi_1\rangle, \tag{4.36}$$

where we have used Eqs. (4.31) and (4.33).[3] Thus, the spin eigenfunction–antisymmetry conditions require that $\psi_1$ and $\psi_2$ have the same normalization, whatever it is. Furthermore, the $P_{ij}$ operators commute with the Hamiltonian of the ESE, and an argument similar to that leading to Eq. (4.36) yields

$$\langle\psi_2|H|\psi_2\rangle = \langle\psi_1|H|\psi_1\rangle. \tag{4.37}$$

These considerations may now be used to simplify the Rayleigh quotient for $\Psi$, and we see that

$$\frac{\langle\Psi|H|\Psi\rangle}{\langle\Psi|\Psi\rangle} = \frac{\langle\psi_1|H|\psi_1\rangle + \langle\psi_2|H|\psi_2\rangle}{\langle\psi_1|\psi_1\rangle + \langle\psi_2|\psi_2\rangle}, \tag{4.38}$$

$$= \frac{\langle\psi_1|H|\psi_1\rangle}{\langle\psi_1|\psi_1\rangle}, \tag{4.39}$$

and contrary to what appeared might be necessary above, we need to determine only one function to obtain the energy. We emphasize that Eq. (4.39) is true even

---

[3] We remind the reader that all permutations are unitary operators. Since binary permutations are equal to their own inverses, they are also Hermitian. Products of commuting binaries are also Hermitian.

for the exact solution to the ESE. The constraints that our spin eigenfunction–antisymmetry conditions impose on the wave function require that $\psi_1$ and $\psi_2$ be closely related, and, if a method is available for obtaining $\psi_1$, $\psi_2$ may then be determined using $P_{ij}$ operators.

If we wish to apply the variation theorem to $\psi_1$, we still need the condition it must satisfy. Reflecting back upon the two-electron systems, we see that the requirement of symmetry for singlet functions could have been written

$$\frac{1}{2}(I + P_{12})\,{}^1\psi(12) = {}^1\psi(12). \tag{4.40}$$

Examining our previous results we see that a corresponding relation for the three-electron case may be constructed:

$$\frac{1}{3}(2I - P_{12}P_{13} - P_{12}P_{23})\,{}^2\psi(123) = {}^2\psi(123). \tag{4.41}$$

This has the correct form: it is Hermitian and idempotent, but that it is actually correct will be more easily ascertained after our general discussion of the next chapter.

## 4.3 Orbital approximation

We now specialize our $\psi$-function, considering it to be a linear combination of products of only three independent orbitals. At the outset we use $a$, $b$, and $c$ to represent three different functions that are to be used as orbitals. To keep the notation from becoming too cumbersome, we use an adaptation of the $[\cdots]$ symbols above. Thus we let

$$[abc] = a(1)b(2)c(3), \tag{4.42}$$
$$[bca] = b(1)c(2)a(3), \tag{4.43}$$

etc. There are, of course, six such functions, since there are six permutations of three objects.

Applying the doublet projector in Eq. (4.41) to each of the six product functions, we obtain the six linear combinations,

$$w_1 = \{2[abc] - [bca] - [cab]\}/3, \tag{4.44}$$
$$w_2 = \{2[bca] - [cab] - [abc]\}/3, \tag{4.45}$$
$$w_3 = \{2[cab] - [abc] - [bca]\}/3, \tag{4.46}$$
$$w_4 = \{2[acb] - [cba] - [bac]\}/3, \tag{4.47}$$
$$w_5 = \{2[cba] - [bac] - [acb]\}/3, \tag{4.48}$$

and

$$w_6 = \{2[bac] - [acb] - [cba]\}/3. \tag{4.49}$$

It is easy to see that these are not all linearly independent: in fact,

$$w_1 + w_2 + w_3 = 0 \tag{4.50}$$

and

$$w_4 + w_5 + w_6 = 0. \tag{4.51}$$

There are, therefore, apparently four functions based upon these orbitals to be used for doublet states. Again, there seems to be too many, but we now show how these are to be used. To proceed, we dispense with $w_3$ and $w_6$, since they are not needed.

We now construct functions that satisfy Eqs. (4.25) and (4.26). By direct calculation we find that

$$P_{12}w_1 = -w_4 - w_5, \tag{4.52}$$
$$P_{12}w_2 = w_5, \tag{4.53}$$
$$P_{12}w_4 = -w_1 - w_2, \tag{4.54}$$

and

$$P_{12}w_5 = w_2. \tag{4.55}$$

The $w$s constitute a basis for a matrix representation of $P_{12}$

$$\mathbf{P}_{12} = \begin{bmatrix} 0 & \mathbf{A} \\ \mathbf{A} & 0 \end{bmatrix}, \tag{4.56}$$

where

$$\mathbf{A} = \begin{bmatrix} -1 & -1 \\ 0 & 1 \end{bmatrix}. \tag{4.57}$$

$\mathbf{A}$ has eigenvalues $\pm 1$ and is diagonalized by the (nonunitary) similarity transformation:

$$\mathbf{M}^{-1}\mathbf{A}\mathbf{M} = \begin{bmatrix} -1 & 0 \\ 0 & 1 \end{bmatrix}, \tag{4.58}$$

$$\mathbf{M} = \begin{bmatrix} 1 & 1 \\ 0 & -2 \end{bmatrix}. \tag{4.59}$$

We now subject $\mathbf{P}_{12}$ to a similarity transformation by $\mathbf{N}$,

$$\mathbf{N} = \begin{bmatrix} \mathbf{M} & -\mathbf{M} \\ \mathbf{M} & \mathbf{M} \end{bmatrix}, \tag{4.60}$$

and obtain

$$N^{-1}P_{12}N = \begin{bmatrix} -1 & 0 & 0 & 0 \\ 0 & 1 & 0 & 0 \\ 0 & 0 & 1 & 0 \\ 0 & 0 & 0 & -1 \end{bmatrix}. \tag{4.61}$$

We may work out $P_{23}$ in the same way and subject it to the same similarity transformation to obtain

$$N^{-1}P_{23}N = \begin{bmatrix} 1/2 & 3/2 & 0 & 0 \\ 1/2 & -1/2 & 0 & 0 \\ 0 & 0 & -1/2 & -3/2 \\ 0 & 0 & -1/2 & 1/2 \end{bmatrix}. \tag{4.62}$$

These do not yet quite satisfy the conditions on antisymmetry given in Eqs. (4.25), (4.26), (4.29), and (4.30), but further transformation by

$$Q = \begin{bmatrix} 1 & 0 & 0 & 0 \\ 0 & -1/\sqrt{3} & 0 & 0 \\ 0 & 0 & 0 & 1 \\ 0 & 0 & 1/\sqrt{3} & 0 \end{bmatrix} \tag{4.63}$$

yields

$$Q^{-1}N^{-1}P_{12}NQ = \begin{bmatrix} -1 & 0 & 0 & 0 \\ 0 & 1 & 0 & 0 \\ 0 & 0 & -1 & 0 \\ 0 & 0 & 0 & 1 \end{bmatrix} \tag{4.64}$$

and

$$Q^{-1}N^{-1}P_{23}NQ = \begin{bmatrix} 1/2 & -\sqrt{3}/2 & 0 & 0 \\ -\sqrt{3}/2 & -1/2 & 0 & 0 \\ 0 & 0 & 1/2 & -\sqrt{3}/2 \\ 0 & 0 & -\sqrt{3}/2 & -1/2 \end{bmatrix}, \tag{4.65}$$

which do agree. Since $P_{13} = P_{12}P_{23}P_{12}$ the requirements for that matrix will also be satisfied. Putting together all of the transformations we eventually arrive at

$$x = Q^{-1}N^{-1}w, \tag{4.66}$$

where $\mathbf{w} = [w_1 \; w_2 \; w_4 \; w_5]^\dagger$ with the components defined in Eqs. (4.44), (4.45), (4.47), and (4.48), respectively. Thus we have the four functions,[4]

$$x_1 = \{[abc] - [cab] + [acb] - [bac]\}/2, \tag{4.67}$$

$$x_2 = \{2[bca] + 2[cba] - [cab] - [acb] - [abc] - [bac]\}/\sqrt{12}, \tag{4.68}$$

$$x_3 = \{2[bca] - 2[cba] - [cab] + [acb] - [abc] + [bac]\}/\sqrt{12}, \tag{4.69}$$

and

$$x_4 = \{-[abc] + [cab] + [acb] - [bac]\}/2. \tag{4.70}$$

Therefore, the four linearly independent functions we obtain in the orbital approximation can be arranged into two pairs of linear combinations, each pair of which satisfies the transformation conditions to give an antisymmetric doublet function. The most general total wave function then requires another linear combination of the pair of functions. In this case Eq. (4.18) can be written

$$^2\Psi = (x_1 + \alpha x_3)^2 \phi_1 + (x_2 + \alpha x_4)^2 \phi_2, \tag{4.71}$$

where $\alpha$ is a new variation parameter that is characteristic of the doublet case when we use orbital product functions. The same value, $\alpha$, is required in both terms because of Eq. (4.35). In addition, Eq. (4.39) is still valid, of course, so that the energy is calculated from

$$W = \frac{\langle (x_1 + \alpha x_3) | H | (x_1 + \alpha x_3) \rangle}{\langle (x_1 + \alpha x_3) | (x_1 + \alpha x_3) \rangle}. \tag{4.72}$$

Thus, even without mixing in configurations of different orbitals, determining the energy of a doublet system of three electrons in three different orbitals is a sort of two-configuration calculation.

The way this function represents the system is strongly influenced by the dynamics of the problem, as well as the flexibility allowed. If we were to find the set of three orbitals and value of $\alpha$ minimizing $W$, we obtain essentially the SCVB wave function. What this looks like depends significantly on the potential energy function. If we are treating the $\pi$ system of the allyl radical, where all three orbitals are nearly degenerate, we obtain one sort of answer. If, on the other hand, we treat a deep narrow potential like the Li atom, we would obtain two orbitals close to one another and like the traditional $1s$ orbital of self-consistent-field (SCF) theory. The third would resemble the $2s$ orbital, of course.

---

[4] These are displayed with an arbitrary overall normalization. This is unimportant in the Rayleigh quotient so long as the functions' normalizations are correct relative to one another. The real normalization constant depends upon the overlaps, of course.

If, as a further approximation, we force the two inner orbitals of the Li-atom treatment to be the same, the need for $\alpha$ disappears. In Eqs. (4.67)–(4.70) we put $c = b$, and the result is

$$x_1 = [abb] - [bab], \tag{4.73}$$
$$x_2 = \{2[bba] - [bab] - [abb]\}/\sqrt{3}, \tag{4.74}$$

and

$$x_3 = x_4 = 0, \tag{4.75}$$

and we no longer have *two* pairs of functions with the correct properties. There is not an extra variation parameter to determine in this case. The most general wave function of this sort with optimized orbitals is the ordinary spin-restricted open-shell Hartree–Fock (ROHF) function.[5]

If we have three orbitals the same, $c = b = a$, we then see that all of the $x_i$ vanish identically. This is clearly the familiar answer: the Pauli exclusion principle prohibits three electrons having the same spatial part of their spin orbitals.

In Chapter 10, after we have discussed the general $n$-electron problem, we will illustrate these two three-electron doublet systems with some calculations. We delay these examples because notational problems will be considerably simpler at that time.

Above, we commented on the unfortunate increase in complexity in going from a two-electron singlet system to a three-electron doublet system. Unfortunately, the complexity accelerates as the number of electrons increases.

---

[5]  The ordinary unrestricted Hartree–Fock (UHF) function is not written like either of these. It is not a pure spin state (doublet) as are these functions. The spin coupled VB (SCVB) function is lower in energy than the UHF in the same basis.

# 5

# Advanced methods for larger molecules

As was seen in the last chapter, the effect of permutations on portions of the wave function is important in enforcing their correct character. The permutations of $n$ entities form a group in the mathematical sense that is said to be one of the *symmetric groups*.[1] In particular, when we have all of the permutations of $n$ entities the group is symbolized $S_n$. In this chapter we give, using the theory of the symmetric groups, a generalization of the special treatment of three electrons discussed above.

There are several more or less equivalent methods for dealing with the twin problems of constructing antisymmetric functions that are also eigenfunctions of the spin. Where orbitals are orthogonal the graphical unitary group approach (GUGA), based upon the symmetric group and unitary group representations, is popular today. With VB functions, which perforce have nonorthogonal orbitals, a significant problem centers around devising algorithms for calculating matrix elements of the Hamiltonian that are efficient enough to be useful. In the past symmetric group methods have been criticized as being overcomplicated. Nevertheless, the present author knows of no other techniques for obtaining what appears to be the optimal algorithm for these calculations.

This chapter is the most complicated and formal in the book. Looking back to Chapter 4 we can obtain an idea of what is needed in general. In this chapter we:

1. outline the theory of the permutation (symmetric) groups and their algebras. The goal here is the special, "factored" form for the antisymmetrizer of Section 5.4.10, since, in this form the influence of the spin state on the spatial functions is especially transparent;
2. show how the resultant spatial functions allow an optimal algorithm for the evaluation of matrix elements of the Hamiltonian, which is given;
3. show the way to generate HLSP functions from the previous treatment.

---

[1] A word of caution here is in order. Groups describing spatial symmetry are frequently spoken of as symmetry groups. These should be distinguished from the *symmetric groups*.

These are the principal ideas of this chapter although the order is not exactly given by the list, and we start with an outline of permutation groups.

## 5.1 Permutations

The word permutation has two meanings in common usage. The standard dictionary definition "an arrangement of a number of objects" is one of them, but we will reserve it to mean *the act of permuting a set of objects*. In this work the word "arrangement" will be used to refer to the particular ordering and "permutation" will always refer to the act of changing the arrangement. The set of "acts" that result in a particular *re*arrangement is not unique, but we do not need to worry about this. We just consider it the permutation producing the rearrangement.

In Chapter 4 we used symbols like $P_{ij}$ to indicate a binary permutation, but this notation is much too inefficient for general use. Another inefficient notation sometimes used is

$$\begin{pmatrix} \downarrow & 1 & 2 & 3 & \cdots & n \\ & i_1 & i_2 & i_3 & \cdots & i_n \end{pmatrix},$$

where $i_1, i_2, \ldots, i_n$ is a different arrangement of the first $n$ integers. We interpret this to mean that the object (currently) in position $j$ is moved to the position $i_j$ (not necessarily different from $j$). The *inverse* permutation could be symbolized by reversing the direction of the arrow to $\uparrow$. There is too much redundancy in this symbol for convenience, and permutations are most frequently written in terms of their *cycle structure*.

Every permutation can be written as a product, in the group sense, of cycles, which are represented by disjoint sets of integers. The symbol (12) represents the interchange of objects 1 and 2 in the set. This is independent of the number of objects.

A cycle of three integers (134) is interpreted as instructions to take the object in position 1 to position 3, that in position 3 to position 4, and that in position 4 to position 1. It should be clear that (134), (341), and (413) all refer to actions with the same result. A permutation may have several cycles, (12)(346)(5789). It should be observed that there are no numbers common between any of the cycles. A unary cycle, e.g., (3), says that the object in position 3 is not moved. In writing permutations unary cycles are normally omitted.

The group nature of the symmetric groups arises because the application of two permutations sequentially is another permutation, and the sequential application can be defined as the group multiplication operation. If we write the product of two permutations,

$$(124) \times (34) = (1243), \tag{5.1}$$

we say (34) is applied first and then (124). Working out results like that of Eq. (5.1) is fairly simple with a little practice. If we decide to write cycles with the smallest number in them first, we would start by searching the product of cycles from the right for the smallest number, which is 1. The rightmost reference to 1 says "1 → 2", the rightmost reference to 2 says "2 → 4", the rightmost reference to 4 says "4 → 3", the rightmost reference to 3 says "3 → 4", but 4 appears again in the left factor where "4 → 1", closing the cycle. If two cycles have no numbers in common, their product is just the two of them written side by side. The order is immaterial; thus they commute. It may be shown also that the product defined this way is associative, $(ab)c = a(bc)$.

The inverse of a cycle is simply obtained by writing the numbers in reverse order. Thus $(1243)^{-1} = (3421) = (1342)$, and $(1243)(1342) = I$, the identity, which corresponds in this case to no action, of course. We have here all the requirements of a finite group.

1. A set of quantities with an associative law of composition yielding another member of the set.
2. An identity appears in the set. The identity commutes with all elements of the set.
3. Corresponding to each member of the set there is an inverse. (The first two laws guarantee that an element commutes with its inverse.)

A cycle can be written as a product of binary permutations in a number of ways. One of these is

$$(i_1 i_2 i_3 \cdots i_{n-1} i_n) = (i_1 i_2)(i_2 i_3) \cdots (i_{n-1} i_n). \tag{5.2}$$

The ternary cycle that is the product of two binary permutations with one number in common can be written in three equivalent ways, $(i_1 i_2)(i_2 i_3) = (i_2 i_3)(i_1 i_3) = (i_1 i_3)(i_1 i_2)$. Clearly, these transformations could be applied to the result of Eq. (5.2) to arrive at a large number of products of different binaries. Nevertheless, each one contains the same number of binary interchanges.

A cycle of $n$ numbers is always the product of $n - 1$ interchanges, regardless of the way it is decomposed. In addition, these decompositions can vary in their *efficiency*. Thus, e.g., $(12)(23)(14)(24)(14) = (23)(13)(12) = (13)$ all represent the same permutation, but they all have an *odd* number of interchanges in their representation.

In general, a permutation is the product of $m_2$ binary cycles, $m_3$ ternary cycles, $m_4$ quaternary cycles, etc., all of which are noninteracting. If all of these are factored into (now interacting) binaries, the number is

$$\sigma = \sum_{j=2}^{j_{max}} (j - 1) m_j, \tag{5.3}$$

which is called the *signature* of the permutation. A permutation is said to be even or odd according to whether $\sigma$ is an even or odd number. The value of $\sigma$ depends upon the efficiency of the decomposition, but its oddness or evenness does not. Therefore, the product of two even or two odd permutations is even, while the product of an even and an odd permutation is odd.

## 5.2 Group algebras

We need to generalize the idea of a group to that of group algebra. The reader has probably already used these ideas without the terminology. The *antisymmetrizer* we have used so much in earlier discussions is just such an entity for a symmetric group, $S_n$,

$$\mathcal{A} = \frac{1}{n!} \sum_{\pi \in S_n} (-1)^{\sigma_\pi} \pi, \tag{5.4}$$

$$= \mathcal{A}^2, \tag{5.5}$$

where $\sigma_\pi$ is the signature of the permutation defined in Eq. (5.3). We note that Eq. (5.4) describes an entity in which we have multiplied group elements by scalars ($\pm 1$) and added the results together. Equation (5.5) implies that we may multiply two such entities together, collect the terms by adding together the coefficients of like permutations, and write the result as an algebra element. Hence, $\mathcal{A}$ is idempotent. NB The assumption that we can identify the individual group elements to collect coefficients is mathematically equivalent to assuming the group elements themselves form a *linearly independent* set of algebra elements.[2] The reader may feel that couching our argument in terms of group algebras is unnecessarily abstract, but, unfortunately, without this idea the arguments become excessively tedious.

Thus, we define the operations of multiplying a group element by a scalar and adding two or more such entities. In this, we assume the elements of the group to be linearly independent, otherwise the mathematical structure we are dealing with would be unworkable. An element, $x$, of the algebra associated with $S_n$ can be written

$$x = \sum_{\rho \in S_n} x_\rho \rho, \tag{5.6}$$

where $x_\rho$ is, in general, a complex number. Two elements of the algebra may be

---

[2] In most arguments involving spatial symmetry, the group character projections used are implicitly (if not explicitly) elements of the algebra of the corresponding symmetry group.

added, subtracted, or multiplied

$$x \pm y = \sum_{\rho \in S_n} (x_\rho \pm y_\rho)\rho \tag{5.7}$$

and

$$xy = \sum_{\rho \in S_n} \sum_{\pi \in S_n} x_\rho y_\pi \, \rho\pi$$

$$= \sum_{\eta \in S_n} \left( \sum_{\rho \in S_n} x_\rho y_{\rho^{-1}\eta} \right) \eta, \tag{5.8}$$

where $\pi = \rho^{-1}\eta$. The way the product is formed in Eq. (5.8) should be carefully noted. We also note that individual elements of the group necessarily possess inverses, but this is not true for the general algebra element.

These considerations make the elements of a group embedded in the algebra behave like a basis for a vector space, and, indeed, this is a normed vector space. Let $x$ be any element of the algebra, and let $[\![x]\!]$ stand for the coefficient of $I$ in $x$. Also, for all of the groups we consider in quantum mechanics it is necessary that the group elements (not algebra elements) are assumed to be unitary. There will be more on this below in Section 5.4 This gives the relation $\rho^\dagger = \rho^{-1}$. Thus we have

$$||x||^2 = [\![x^\dagger x]\!] = \sum_\rho |x_\rho|^2 \geq 0, \tag{5.9}$$

where the equality holds if and only if $x = 0$. One of the important properties of $[\![xy]\!]$ is

$$[\![xy]\!] = [\![yx]\!] \tag{5.10}$$

for any two elements of the algebra. We will frequently use the "$[\![\cdots]\!]$" notation in later work.

Since the group elements we are working with normally arise as operators on wave functions in quantum mechanical arguments, by extension, the algebra elements also behave this way. Because of the above, one of the important properties of their manipulation is

$$\langle \phi | x\psi \rangle = \langle x^\dagger \phi | \psi \rangle. \tag{5.11}$$

The idea of a group algebra is very powerful and allowed Frobenius to show constructively the entire structure of irreducible matrix representations of finite groups. The theory is outlined by Littlewood[37], who gives references to Frobenius's work.

## 5.3 Some general results for finite groups

### 5.3.1 Irreducible matrix representations

Many works[5, 6] on group theory describe matrix representations of groups. That is, we have a set of matrices, one for each element of a group, $G$, that satisfies

$$D(\rho)D(\eta) = D(\rho\eta), \tag{5.12}$$

for each pair of group elements.[3] Any matrix representation may be subjected to a similarity transformation to obtain an equivalent representation:

$$\bar{D}(\rho) = N^{-1}D(\rho)N; \qquad \rho \in G, \tag{5.13}$$

where $N$ is any nonsingular matrix. Amongst all of the representations, unitary ones are frequently singled out. This means that

$$D(\rho)^{-1} = D(\rho^{-1}) = D(\rho)^{\dagger}, \tag{5.14}$$

and, for a finite group, such unitarity is always possible to arrange. For our work, however, we need to consider representations that are not unitary, so some of the results quoted below will appear slightly different from those seen in expositions where the unitary property is always assumed.

The theory of group representation proves a number of results.

1. There is a set of inequivalent irreducible representations. The number of these is equal to the number of equivalence classes among the group elements. If the $\alpha^{\text{th}}$ irreducible representation is an $f_\alpha \times f_\alpha$ matrix, then

$$\sum_\alpha f_\alpha^2 = g, \tag{5.15}$$

   where $g$ is the number of elements in the group.

2. The elements of the irreducible representation matrices satisfy a somewhat complicated law of composition:

$$\sum_\rho D_{ji}^\alpha(\rho^{-1})D_{lk}^\beta(\eta\rho) = \delta_{\alpha\beta}\delta_{jk}\frac{g}{f_\alpha}D_{li}^\alpha(\eta). \tag{5.16}$$

3. If we specify in the previous item that $\eta = I$, the *orthogonality theorem* results:

$$\sum_\rho D_{ji}^\alpha(\rho^{-1})D_{lk}^\beta(\rho) = \delta_{\alpha\beta}\delta_{jk}\delta_{il}\frac{g}{f_\alpha}. \tag{5.17}$$

Equation (5.17) has an important implication. Consider a large table with entries, $D_{ij}^\alpha(\rho^{-1})$, and the rows labeled by $\rho$ and the columns labeled by the possible values of $\alpha, i,$ and $j$. Because of Eq. (5.15) the table is square, and may be considered

---

[3] The representation property does not imply, however, that $\rho \neq \eta$ implies $D(\rho) \neq D(\eta)$.

a $g \times g$ matrix. Because of Eq. (5.17) the matrix is necessarily nonsingular and possesses an inverse. The inverse matrix is then an array with $g$ rows and $g$ columns where the entries[4] are $(f_\alpha/g)D^\alpha_{ji}(\rho)$, the rows are labeled by $\alpha$, $i$, and $j$, and the columns by $\rho$. In the theory of matrices, it may be proved that matrix inverses commute, therefore we have another relation among the irreducible representation matrix elements:

$$\sum_{\alpha ij} \frac{f_\alpha}{g} D^\alpha_{ji}(\eta) D^\alpha_{ij}(\rho^{-1}) = \delta_{\eta\rho}, \qquad (5.18)$$

where $\delta_{\eta\rho}$ is 1 or 0, according as $\eta$ and $\rho$ are or are not the same.

### 5.3.2 Bases for group algebras

The matrices of the irreducible representations provide one with a special set of group algebra elements. We define

$$e^\alpha_{ij} = \frac{f_\alpha}{g} \sum_\rho D^\alpha_{ji}(\rho^{-1})\rho, \qquad (5.19)$$

and using Eq. (5.16) one can show that

$$e^\alpha_{ij} e^\beta_{kl} = \delta_{\alpha\beta} \delta_{jk} e^\alpha_{il}. \qquad (5.20)$$

Equation (5.19) gives the algebra basis as a sum over the group elements. Using Eq. (5.18) we may also write the group elements as a sum over the algebra basis,

$$\rho = \sum_{\alpha ij} D^\alpha_{ij}(\rho) e^\alpha_{ij} \qquad (5.21)$$

and, if $\rho$ is taken as the identity,

$$\sum_{\alpha i} e^\alpha_{ii} = I. \qquad (5.22)$$

In the theory of operators over vector spaces Eq. (5.22) is said to give the *resolution of the identity*, since by Eq. (5.20) each $(e^\alpha_{ii})^2 = e^\alpha_{ii}$, and is *idempotent*.

We note another important property of these bases. Irreducible representation matrices may be obtained from the $e^\alpha_{ij}$ by using the relation

$$D^\alpha_{ij}(\rho) = [\![ \rho e^\alpha_{ji} ]\!], \qquad (5.23)$$

where we have used the $[\![ \cdots ]\!]$ notation defined above to obtain the coefficient of the identity operation.

---

[4] NB In the inverse we have interchanged the index labels of the irreducible representation matrix.

The theory of matrix representations of groups is more commonly discussed than the theory of group algebras. The latter are, however, important for our discussion of the symmetric groups, because Young (this theory is discussed by Rutherford[7]) has shown, for these groups, how to generate the algebra first and obtain the matrix representations from them. In fact, we need not obtain the irreducible representation matrices at all for our work; the algebra elements *are* the operators we need to construct spatial VB basis functions appropriate for a given spin.

## 5.4 Algebras of symmetric groups

### 5.4.1 The unitarity of permutations

Before we actually take up the subject of this section we must give a demonstration that permutations are unitary. This was deferred from above.

The $n$-particle spatial (or spin) functions we work with are elements of a Hilbert space in which the permutations are operators. If $\Xi(1, 2, \ldots, n)$ and $\Upsilon(1, 2, \ldots, n)$ are two such functions we generally understand that

$$\langle \Xi | \Upsilon \rangle \equiv \int \Xi(1, 2, \ldots, n)^* \Upsilon(1, 2, \ldots, n)\, d\tau_1 d\tau_2 \cdots d\tau_n. \tag{5.24}$$

If $P_{op}$ and $Q_{op}$ are operators in the Hilbert space and

$$\langle Q_{op} \Xi | \Upsilon \rangle = \langle \Xi | P_{op} \Upsilon \rangle \tag{5.25}$$

for all $\Xi$ and $\Upsilon$ in the Hilbert space, $Q_{op}$ is said to be the *Hermitian conjugate* of $P_{op}$, i.e., $Q_{op} = P_{op}^\dagger$. Consider the integral

$$\int \Xi(1, 2, \ldots, n)^* \pi \, \Upsilon(1, 2, \ldots, n)\, d\tau_1 d\tau_2 \cdots d\tau_n$$

$$= \int [\pi^{-1} \Xi(1, 2, \ldots, n)]^* \Upsilon(1, 2, \ldots, n)\, d\tau_1 d\tau_2 \cdots d\tau_n, \tag{5.26}$$

where $\pi$ is some permutation. Equation (5.26) follows because of the possibility of relabeling variables of definite integrals, and, since it is true for all $\Xi$ and $\Upsilon$,

$$\pi^\dagger = \pi^{-1}. \tag{5.27}$$

This is the definition of a unitary operator.

### 5.4.2 Partitions

The theory of representations of symmetric groups is intimately connected with the idea of partitions of integers. Rutherford[7] gives what is probably the most accessible treatment of these matters. A *partition* of an integer $n$ is a set of smaller integers, not necessarily different, that add to $n$. Thus $5 = 3 + 1 + 1$ constitutes a partition of 5. Partitions are normally written in $\{\cdots\}$, and another way of writing the partition of 5 is $\{3,1^2\}$. We use exponents to indicate multiple occurrences of numbers in the partition, and we will write them with the numbers in decreasing order. The distinct partitions of 5 are $\{5\}$, $\{4,1\}$, $\{3,2\}$, $\{3,1^2\}$, $\{2^2,1\}$, $\{2,1^3\}$, and $\{1^5\}$. There are seven of them, and the theory of the symmetric groups says that this is also the number of inequivalent irreducible representations for the group $S_5$ made up of all the permutations of five objects. We have written the above partitions of 5 in the standard order, such that partition $i$ is before partition $j$ if the first number in $i$ differing from the corresponding one in $j$ is larger than the one in $j$. When we wish to speak of a general partition, we will use the symbol, $\lambda$.

### 5.4.3 *Young tableaux and $\mathcal{N}$ and $\mathcal{P}$ operators*

Associated with each partition there is a table, called by Young a *tableau*. In our example using 5, we might place the integers 1 through 5 in a number of rows corresponding to the integers in a partition, each row having the number of entries of that part of the partition, e.g., for $\{3,2\}$ and $\{2^2,1\}$ we would have

$$\begin{bmatrix} 1 & 2 & 3 \\ 4 & 5 & \end{bmatrix} \quad \text{and} \quad \begin{bmatrix} 1 & 2 \\ 3 & 4 \\ 5 & \end{bmatrix},$$

respectively. The integers might be placed in another order, but, for now, we assume they are in sequential order across the rows, finishing each row before starting the next.

Associated with each tableau, we may construct two elements of the group algebra of the corresponding symmetric group. The first of these is called the *row symmetrizer* and is symbolized by $\mathcal{P}$. Each row of the tableau consists of a distinct subset of the integers from 1 through $n$. If we add together all of the permutations involving just those integers in a row with the identity, we obtain the symmetrizer for that row. Thus for the $\{3,2\}$ tableau, the symmetrizer for the first row is

$$I + (12) + (13) + (23) + (123) + (132)$$

and for the second is

$$I + (45).$$

Thus, the total row symmetrizer, $\mathcal{P}$ is

$$\mathcal{P} = [I + (12) + (13) + (23) + (123) + (132)][I + (45)]. \tag{5.28}$$

The second of these is the *column antisymmetrizer* and is symbolized by $\mathcal{N}$. As might be expected, for the {3,2} tableau the column antisymmetrizer is the product of the antisymmetrizer for each column and is

$$\mathcal{N} = [I - (14)][I - (25)]. \tag{5.29}$$

In these expressions a symmetrizer is the sum of all of the corresponding permutations and the antisymmetrizer is the sum with plus signs for even permutations and minus signs for odd permutations. In Eqs. (5.28) and (5.29) the specific $\mathcal{P}$ and $\mathcal{N}$ are given for the arrangement of numbers in the {3,2} tableau above. A different arrangement of integers in this same shape would in many, but not all, cases give different $\mathcal{P}$ and $\mathcal{N}$ operators.

As a further example we give the $\mathcal{P}$ and $\mathcal{N}$ operators for the above tableau associated with the shape $\{2^2,1\}$. For this we have

$$\mathcal{P} = [I + (12)][I + (34)],$$
$$\mathcal{N} = [I - (13) - (15) - (35) + (135) + (153)][I - (24)].$$

Here, again, $I$ is the only operation in common between $\mathcal{P}$ and $\mathcal{N}$.

A central result of Young's theory is that the product $\mathcal{N}\mathcal{P}$ is proportional to an algebra element that will serve as one of the $e_{ii}^{\alpha}$ basis elements discussed above, and the proportionality constant is $f_{\alpha}/n!$, $n!$ being the value of $g$ in this case. The product $\mathcal{P}\mathcal{N}$ serves equally well, but is, of course, a different element of the algebra, since $\mathcal{N}$ and $\mathcal{P}$ do not normally commute.

### 5.4.4 Standard tableaux

In a tableau corresponding to a partition of $n$, there are, of course, $n!$ different arrangements of the way the first $n$ integers may be entered. Among these there is a subset that Young called *standard tableaux*. These are those for which the numbers in any row increase to the right and downward in any column. Thus, we have for {3,2}

$$\begin{bmatrix} 1 & 2 & 3 \\ 4 & 5 & \end{bmatrix}, \begin{bmatrix} 1 & 2 & 4 \\ 3 & 5 & \end{bmatrix}, \begin{bmatrix} 1 & 2 & 5 \\ 3 & 4 & \end{bmatrix}, \begin{bmatrix} 1 & 3 & 4 \\ 2 & 5 & \end{bmatrix}, \text{ and } \begin{bmatrix} 1 & 3 & 5 \\ 2 & 4 & \end{bmatrix},$$

and among the 120 possible arrangements, only five are standard tableaux. These standard tableaux have been ordered in a particular way called a *lexical sequence*. We label the standard tableaux, $T_1, T_2, \ldots$ and imagine the numbers of the tableau written out in a line, row 1, row 2, .... We say that $T_i$ is before $T_j$ if the first number of $T_j$ that differs from the corresponding one in $T_i$ is the larger of the two. In our succeeding work we express the idea of $T_i$ being earlier than $T_j$ with the symbols $T_i < T_j$.

The operators $\mathcal{P}_i$ and $\mathcal{N}_i$ corresponding to $T_i$ are, of course, different and, in fact, have no permutations in common other than the identity. The first important result is that

$$[\![\mathcal{N}_i\mathcal{P}_i]\!] = 1, \tag{5.30}$$

since the only permutation $\mathcal{N}_i$ and $\mathcal{P}_i$ have in common is $I$, and the numbers adding to the coefficient of the identity cannot cancel. Thus $\mathcal{N}_i\mathcal{P}_i$ is never zero.

The second important result here is that

$$\mathcal{P}_i\mathcal{N}_j = \mathcal{N}_j\mathcal{P}_i = 0, \quad \text{if } T_i < T_j. \tag{5.31}$$

This is so because there is some pair of numbers appearing in the same row of $T_i$ that must appear in the same column of $T_j$, if it is later. To see this suppose that the entries in the tableaux are $(T_i)_{kl}$ and $(T_j)_{kl}$, where $k$ and $l$ designate the row and column in the shape. Let the first difference occur at row $m$ and column $n$. Thus, $(T_j)_{mn} > (T_i)_{mn}$, but $(T_i)_{mn}$ must appear somewhere in $T_j$. Because of the way standard tableaux are ordered it must be $(T_j)_{m'n'}$, where $m' > m$ and $n' < n$. Now, also by hypothesis, $(T_j)_{mn'} = (T_i)_{mn'}$, since this is in the region where the two are the same. Therefore, there is a pair of numbers in the same row of $T_i$ that appear in the same column of $T_j$. Calling these numbers $p$ and $q$, we have

$$\mathcal{P}_i = (1/2)\mathcal{P}_i[I + (pq)], \tag{5.32}$$
$$= (1/2)[I + (pq)]\mathcal{P}_i, \tag{5.33}$$
$$\mathcal{N}_j = (1/2)\mathcal{N}_j[I - (pq)], \tag{5.34}$$
$$= (1/2)[I - (pq)]\mathcal{N}_j, \tag{5.35}$$

and

$$[I + (pq)][I - (pq)] = [I - (pq)][I + (pq)] = 0. \tag{5.36}$$

One should not conclude, however, that $\mathcal{P}_j\mathcal{N}_i = \mathcal{N}_i\mathcal{P}_j = 0$ if $T_i < T_j$. Although true in some cases, we see that it does not hold true for the first and last of the tableaux above. No pair in a row of the last is in a column of the first. In fact, the nonstandard tableau

$$T' = \begin{bmatrix} 1 & 5 & 3 \\ 4 & 2 & \end{bmatrix}$$

can be obtained from $T_5$ by permutations within rows and from $T_1$ by permutations within columns. Thus, $\mathcal{P}' = \mathcal{P}_5$ and $\mathcal{N}' = \pm\mathcal{N}_1$, and, therefore,

$$\mathcal{P}_5\mathcal{N}_1 = \pm\mathcal{P}'\mathcal{N}' \neq 0. \tag{5.37}$$

This would also be true for the operators written in the other order.

We stated above that there is an inequivalent irreducible representation of $S_n$ associated with each partition of $n$, and we use the symbol $f_\lambda$ to represent the number of standard tableaux corresponding to the partition, $\lambda$. Using induction on $n$, Young proved the theorem

$$\sum_\lambda f_\lambda^2 = n!,  \tag{5.38}$$

which should be compared with Eq. (5.15).

Young also derived a formula for $f_\lambda$, but, as will be seen, we need only a small number of partitions for our work with fermions like electrons. These are either $\{n - k, k\}$ or $\{2^k, 1^{n-2k}\}$ for all $k = 0, 1, \ldots$, such that $n - 2k \geq 0$. In fact, the shapes of the tableaux corresponding to these two partitions are closely related, being *transposes* of one another. Letting $n = 5$ and $k = 2$, the shape of $\{3,2\}$ may be symbolized with dots as

$$\begin{array}{ccc} \bullet & \bullet & \bullet \\ \bullet & \bullet & \end{array}$$

If we interchange rows and columns in this shape, we obtain

$$\begin{array}{cc} \bullet & \bullet \\ \bullet & \bullet \\ \bullet & \end{array}$$

which is seen to be the shape of the partition $\{2^2, 1\}$. Partition shapes and tableaux related this way are said to be *conjugates*, and we use the symbol $\tilde{\lambda}$ to represent the partition conjugate to $\lambda$.

It should be reasonably self-evident that the conjugate of a standard tableau is a standard tableau of the conjugate shape. Therefore, $f_\lambda = f_{\tilde{\lambda}}$, and irreducible representations corresponding to conjugate partitions are the same size. In fact, the irreducible representations are closely related. If $D^\lambda(\rho)$ is one of the irreducible representation matrices for partition $\lambda$, one has

$$D^{\tilde{\lambda}}(\rho) = (-1)^{\sigma_\rho} D^\lambda(\rho),  \tag{5.39}$$

where $\sigma_\rho$ is the signature of $\rho$.

As we noted above, Young derived a general expression for $f_\lambda$ for any shape. For the partitions we need there is, however, some simplification of the general expression, and we have for either $\{n - k, k\}$ or $\{2^k, 1^{n-2k}\}$

$$f_\lambda = \frac{n - 2k + 1}{n + 1} \binom{n + 1}{k},  \tag{5.40}$$

$$\binom{p}{q} = \frac{p!}{q!(p - q)!}.  \tag{5.41}$$

### 5.4.5 The linear independence of $\mathcal{N}_i \mathcal{P}_i$ and $\mathcal{P}_i \mathcal{N}_i$

The relationships expressed in Eq. (5.31) can be used to prove the very important result that the set of algebra elements, $\mathcal{N}_i \mathcal{P}_i$, is linearly independent. First, from Eq. (5.30) we have seen that they are not zero, so we suppose there is a relation

$$\sum_i a_i \mathcal{N}_i \mathcal{P}_i = 0. \tag{5.42}$$

We multiply Eq. (5.42) on the right, starting with the final one $\mathcal{N}_f$, and, because of Eq. (5.31), we obtain

$$a_f \mathcal{N}_f \mathcal{P}_f \mathcal{N}_f = 0. \tag{5.43}$$

Therefore, either $a_f$ or $\mathcal{N}_f \mathcal{P}_f \mathcal{N}_f$, or both must be 0. We observe, however, that

$$[\![\mathcal{N}_i \mathcal{P}_i \mathcal{N}_i]\!] = [\![\mathcal{N}_i^2 \mathcal{P}_i]\!] \tag{5.44}$$

$$= g_\mathcal{N} [\![\mathcal{N}_i \mathcal{P}_i]\!] \tag{5.45}$$

$$= g_\mathcal{N}, \tag{5.46}$$

where $g_\mathcal{N}$ is the order of the subgroup of $\mathcal{N}$, and this is true for any $i$. Thus, $\mathcal{N}_f \mathcal{P}_f \mathcal{N}_f$ is not zero and $a_f$ in Eq. (5.43) must be. Now that we know $a_f$ is zero, we may multiply Eq. (5.42) on the right by $\mathcal{N}_{f-1}$, and see that $a_{f-1}$ must also be zero. Proceeding this way until we reach $\mathcal{N}_1$, we see that all of the $a_i$ are zero, and the result is proved.

Permutations are unitary operators as seen in Eq. (5.27). This tells us how to take the Hermitian conjugate of an element of the group algebra,

$$x^\dagger = \left( \sum_\pi x_\pi \pi \right)^\dagger, \tag{5.47}$$

$$= \sum_\pi x_\pi^* \pi^{-1}, \tag{5.48}$$

$$= \sum_\pi x_{\pi^{-1}}^* \pi. \tag{5.49}$$

In passing we note that $\mathcal{N}$ and $\mathcal{P}$ are Hermitian, since the coefficients are real and equal for inverse permutations.

In general $\mathcal{P}_i \mathcal{N}_i$ is not equal to $\mathcal{N}_i \mathcal{P}_i$ but is its Hermitian conjugate, since $(\rho\pi)^\dagger = \pi^\dagger \rho^\dagger$. Therefore, it should be reasonably obvious that the $\mathcal{P}_i \mathcal{N}_i$ operators are also linearly independent. We note that an alternative, but very similar, proof that all $a_i = 0$ in Eq. (5.42) could be constructed by multiplying on the left by $\mathcal{P}_j$; $j = 1, 2, \ldots, f$ sequentially.

It is now fairly easy to see that we could form a new set of linearly independent quantities

$$x_i \mathcal{N}_i \mathcal{P}_i; \qquad i = 1, 2, \ldots, f, \tag{5.50}$$

where $x_i$ is any set of elements of the algebra that do not result in $x_i \mathcal{N}_i \mathcal{P}_i = 0$. Corresponding results are true for right multiplication (i.e., $\mathcal{N}_i \mathcal{P}_i x_i$). As is probably not surprising there are parallel results for right or left multiplication on $\mathcal{P}_i \mathcal{N}_i$. An important application of this result (for left multiplication) is an algebra element like $x_i = \rho \mathcal{P}_i$, where $\rho$ is any operation of the group, with corresponding expressions for the other cases.

The operators $\mathcal{P}_i$ and $\mathcal{P}_j$ differ only in being based upon a different arrangement of the numbers in the standard tableau they are associated with. Therefore, there exists a permutation, $\pi_{ij}$ that will interconvert $\mathcal{P}_i$ and $\mathcal{P}_j$ with the relation

$$\pi_{ij}\mathcal{P}_j = \mathcal{P}_i \pi_{ij}, \tag{5.51}$$

with a similar expression for $\mathcal{N}_i$ and $\mathcal{N}_j$. The theorems of this section can thus be stated in a different way. For example, we see that the quantities, $\mathcal{P}_1 \mathcal{N}_1 \pi_{1j} = \pi_{1j} \mathcal{P}_j \mathcal{N}_j$, satisfy the definition of Eq. (5.50), and are thus linearly independent. Three similar results pertain for the other three possible combinations of the ordering of the products of $\mathcal{P}$ and $\mathcal{N}$ on either side of the equation. Explicitly, for one of these cases, we may write that the relation

$$\sum_i \mathcal{P}_1 \mathcal{N}_1 \pi_{1i} a_i = 0 \tag{5.52}$$

implies that all $a_i = 0$, with similar implications for the other cases.

### 5.4.6 Von Neumann's theorem

Von Neumann proved a very useful theorem for our work (quoted by Rutherford[7]). Using our notation it can be written

$$\mathcal{P}x\mathcal{N} = [\![\mathcal{P}x\mathcal{N}]\!]\mathcal{P}\mathcal{N}, \tag{5.53}$$

where $x$ is any element of the algebra and $\mathcal{N}$ and $\mathcal{P}$ are based upon the same tableau. A similar expression holds for $\mathcal{N}x\mathcal{P}$.

### 5.4.7 Two Hermitian idempotents of the group algebra

We will choose arbitrarily to work with the first of the standard tableaux[5] of a given partition. With this we can form the two *Hermitian* algebra elements

$$u = \theta \mathcal{P} \mathcal{N} \mathcal{P} \tag{5.54}$$

---

[5] Any tableau would do, but we only need one. This choice serves.

and

$$u' = \theta' \mathcal{N} \mathcal{P} \mathcal{N}, \tag{5.55}$$

where $\theta$ and $\theta'$ are real. We must work out what to set these values to so that $u^2 = u$ and $u'^2 = u'$.

We stated at the end of Section 5.4.3 that $(f/g)\mathcal{N}\mathcal{P}$ or $(f/g)\mathcal{P}\mathcal{N}$, $g = n!$, will serve as an idempotent element of the algebra associated with the partition upon which they are based, although these are, of course, not Hermitian. This means that

$$\mathcal{N}\mathcal{P}\mathcal{N}\mathcal{P} = \frac{g}{f}\mathcal{N}\mathcal{P}. \tag{5.56}$$

Thus, observing that $\mathcal{P}^2 = g_\mathcal{P}\mathcal{P}$, we have

$$(\mathcal{P}\mathcal{N}\mathcal{P})^2 = \mathcal{P}\mathcal{N}\mathcal{P}^2\mathcal{N}\mathcal{P}, \tag{5.57}$$
$$= g_\mathcal{P}\mathcal{P}\mathcal{N}\mathcal{P}\mathcal{N}\mathcal{P}, \tag{5.58}$$
$$= \frac{g g_\mathcal{P}}{f}\mathcal{P}\mathcal{N}\mathcal{P}, \tag{5.59}$$

where $g_\mathcal{P}$ is the order of the subgroup of the $\mathcal{P}$ operator. Thus, we obtain

$$u = \frac{f}{g g_\mathcal{P}}\mathcal{P}\mathcal{N}\mathcal{P} \tag{5.60}$$

as an idempotent of the algebra that is Hermitian. A very similar analysis gives

$$u' = \frac{f}{g g_\mathcal{N}}\mathcal{N}\mathcal{P}\mathcal{N}, \tag{5.61}$$

where $g_\mathcal{N}$ is the order of the subgroup of the $\mathcal{N}$ operator. Although portions of the following analysis could be done with the original non-Hermitian Young idempotents, the operators of Eqs. (5.60) and (5.61) are required near the end of the theory and, indeed, simplify many of the intervening steps.

### 5.4.8 A matrix basis for group algebras of symmetric groups

In the present section we will give a construction of the matrix basis only for the $u = \theta\mathcal{P}\mathcal{N}\mathcal{P}$ operator. The treatment for the other Hermitian operator above is identical and may be supplied by the reader.

Consider now the quantities,

$$m_{ij} = \pi_{i1}u\pi_{1j}, \tag{5.62}$$
$$= (m_{ji})^\dagger, \tag{5.63}$$
$$= \pi_i^{-1}u\pi_j; \quad \pi_j = \pi_{1j}, \tag{5.64}$$

none of which is zero. Since u is based upon the first standard tableau, from now on we suppress the "1" subscript in these equations. This requires us, however, to use the inverse symbol, as seen. Now familiar methods are easily used to show that $m_{ij} \neq 0$ for all $i$ and $j$. In fact, the above results show that the $m_{ij}$ constitute $f^2$ linearly independent elements of the group algebra that, because of Young's results, completely span the space associated with the irreducible representation labeled with the partition. Thus, because of Eq. (5.38) we have found a complete set of linearly independent elements of the whole group algebra.

We now determine the multiplication rule for $m_{ij}$ and $m_{kl}$,

$$m_{ij}m_{kl} = \pi_i^{-1}u\pi_j\pi_k^{-1}u\pi_l. \tag{5.65}$$

Examining the inner factors of this product, we see that

$$u\pi_j\pi_k^{-1}u = \theta^2 \mathcal{PNP}\pi_j\pi_k^{-1}\mathcal{PNP}, \tag{5.66}$$

$$\theta = \frac{f}{gg_P}. \tag{5.67}$$

We now apply Eq. (5.53) to some inner factors and obtain

$$\mathcal{P}\pi_j\pi_k^{-1}\mathcal{PN} = [\![\mathcal{P}\pi_j\pi_k^{-1}\mathcal{PN}]\!]\mathcal{PN}, \tag{5.68}$$

$$= [\![\pi_j\pi_k^{-1}\mathcal{PNP}]\!]\mathcal{PN}, \tag{5.69}$$

$$u\pi_j\pi_k^{-1}u = \theta^2[\![\pi_j\pi_k^{-1}\mathcal{PNP}]\!]\mathcal{PNPNP}, \tag{5.70}$$

$$= \theta^2 g_P^{-1}[\![\pi_j\pi_k^{-1}\mathcal{PNP}]\!]\mathcal{PNPPNP}, \tag{5.71}$$

$$= M_{kj}u, \tag{5.72}$$

$$M_{kj} = g_P^{-1}[\![\pi_k^{-1}\mathcal{PNP}\pi_j]\!]. \tag{5.73}$$

Putting these transformations together,

$$m_{ij}m_{kl} = M_{kj}m_{il}. \tag{5.74}$$

All of the coefficients in $\mathcal{PNP}$ are real and the matrix $M$ is thus real symmetric (and Hermitian). Since the $m_{ij}$ are linearly independent, $M$ must be nonsingular. In addition, $g_P^{-1}[\![\mathcal{PNP}]\!]$ is equal to 1, so the diagonal elements of $M$ are all 1. $M$ is essentially an overlap matrix due to the non-orthogonality of the $m_{ij}$.

We note that if the matrix $M$ were the identity, the $m_{ij}$ would satisfy Eq. (5.20). An orthogonalization transformation of $M$ may easily be effected by the nonsingular matrix $N$

$$N^\dagger M N = I, \tag{5.75}$$

as we saw in Section 1.4.2. It should be recalled that $N$ is not unique; added conditions are required to make it so. We wish $m_{11}$ to be unchanged by the transformation, and an upper triangular $N$ will accomplish both of these goals. If we require all of the diagonal elements of $N$ also to be positive, it becomes uniquely determined. Making the transformations we have

$$e_{ij} = \sum_{kl}(N^\dagger)_{ik}N_{lj}m_{kl}, \tag{5.76}$$

$$e_{ij}e_{kl} = \delta_{jk}e_{il}, \tag{5.77}$$

$$e_{11} = m_{11}, \tag{5.78}$$

as desired. These $e_{ij}$s constitute a real matrix basis for the symmetric group and, clearly, generate a real unitary representation through the use of Eq. (5.23).

### 5.4.9 Sandwich representations

The reader might ask: "Is there a parallel to Eq. (5.23) for the nonorthogonal matrix basis we have just described?" We answer this in the affirmative and show the results.

Clearly, we can define matrices

$$T(\rho)_{ij} = [\![\rho m_{ij}]\!], \tag{5.79}$$

and it is seen that a normal unitary representation may be obtained from

$$D(\rho)_{ij} = \sum_{kl}(N^\dagger)_{ik}T(\rho)_{kl}N_{lj}, \tag{5.80}$$

where we have used Eq. (5.76). The upshot of these considerations is that the $T(\rho)$ matrices satisfy

$$T(\rho)M^{-1}T(\pi) = T(\rho\pi), \tag{5.81}$$

and these have been called *sandwich representations*, because of a fairly obvious analogy. In arriving at Eq. (5.81) we have used

$$NN^\dagger = M^{-1}, \tag{5.82}$$

which is a consequence of Eq. (5.75).

We may also derive a result analogous to Eq. (5.21),

$$\rho = \sum_{\lambda ijkl}(M^{-1})^\lambda_{ki}T(\rho)^\lambda_{kl}(M^{-1})^\lambda_{lj}m^\lambda_{ij}, \tag{5.83}$$

where we have added a partition label to each of our matrices and summed over it.

### *5.4.10 Group algebraic representation of the antisymmetrizer*

As we have seen in Eq. (5.21), an element of the group may be written as a sum over the algebra basis. For the symmetric groups, this takes the form,

$$\rho = \sum_{\lambda ij} D^{\lambda}_{ij}(\rho) e^{\lambda}_{ij}. \tag{5.84}$$

We wish to apply permutations, and the antisymmetrizer to products of spin-orbitals that provide a basis for a variational calculation. If each of these represents a pure spin state, the function may be factored into a spatial and a spin part. Therefore, the whole product, $\Psi$, may be written as a product of a separate spatial function and a spin function. Each of these is, of course, a product of spatial or spin functions of the individual particles,

$$\Psi = \Xi \Theta_{M_s}, \tag{5.85}$$

where $\Xi$ is a product of orbitals and $\Theta_{M_s}$ is a sum of products of spin functions that is an eigenfunction of the total spin. It should be emphasized that the spin function has a definite $M_s$ value, as indicated. If we apply a permutation to $\Psi$, we are really applying the permutation separately to the space and spin parts, and we write

$$\rho \Psi = \rho_r \Xi \rho_s \Theta_{M_s}, \tag{5.86}$$

where the $r$ or $s$ subscripts indicate permutations affecting spatial or spin functions, respectively. Since we are defining permutations that affect only one type of function, separate algebra elements also arise: $e^{\lambda}_{ij,r}$ and $e^{\lambda}_{ij,s}$. These considerations provide us with a special representation of the antisymmetrizer[6] that is useful for our purposes:

$$\mathcal{A} = \frac{1}{g} \sum_{\rho \in S_n} (-1)^{\sigma_{\rho}} \rho_r \rho_s \tag{5.87}$$

$$= \sum_{\rho} \sum_{\lambda ij} \sum_{\lambda' i' j'} D^{\lambda}_{ij}(\rho) e^{\lambda}_{ij,r} D^{\tilde{\lambda}'}_{i'j'}(\rho) e^{\lambda'}_{i'j',s} \tag{5.88}$$

$$= \sum_{\lambda ij} \frac{1}{f_{\lambda}} e^{\lambda}_{ij,r} e^{\tilde{\lambda}}_{ij,s}, \tag{5.89}$$

where we have used Eq. (5.18) and the symbol for the conjugate partition.

In line with the last section we give a version of Eq. (5.89) using the non-orthogonal matrix basis,

$$\mathcal{A} = \sum_{\lambda ijkl} \frac{1}{f_{\lambda}} (M^{-1})^{\lambda}_{ij} m^{\lambda}_{jk,r} (M^{-1})^{\lambda}_{kl} m^{\tilde{\lambda}}_{il,s}, \tag{5.90}$$

where we need not distinguish between the $M^{-1}$ matrices for conjugate partitions.

---

[6] We use the antisymmetrizer in its idempotent form rather than that with the $(\sqrt{n!})^{-1}$ prefactor.

## 5.5 Antisymmetric eigenfunctions of the spin

In this section we investigate the connections between the symmetric groups and spin eigenfunctions. We have briefly outlined properties of spin operators in Section 4.1. The reader may wish to review the material there.

One of the important properties of all of the spin operators is that they are *symmetric*. The *total* vector spin operator is a sum of the vector operators for individual electrons

$$\vec{S} = \sum_{i=1}^{n} \vec{S}_i, \tag{5.91}$$

indicating that the electrons are being treated *equivalently* in these expressions. This means that every $\pi \in S_n$ must commute with the total vector spin operator. Since all of the other operators, $\vec{S}^2$, raising, and lowering operators, are algebraic functions of the components of $\vec{S}$, they also commute with every permutation. We use this result heavily below.

### 5.5.1 Two simple eigenfunctions of the spin

Consider an $n$ electron system in a pure spin state $S$. The associated partition is $\{n/2 + S, n/2 - S\}$, and the first standard tableau is

$$\begin{bmatrix} 1 & \cdots & n/2 - S & \cdots & n/2 + S \\ n/2 + S + 1 & \cdots & n & \end{bmatrix},$$

where we have written the partition in terms of the $S$ quantum number we have targeted. We consider also an array of individual spin functions with the same shape and all $\eta_{1/2}$ in the first row and $\eta_{-1/2}$ in the second

$$\begin{bmatrix} \alpha & \cdots & \alpha & \cdots & \alpha \\ \beta & \cdots & \beta & \end{bmatrix},$$

where we have used the common abbreviations $\alpha = \eta_{1/2}$ and $\beta = \eta_{-1/2}$. Associating symbols in corresponding positions of these two graphical shapes generates a product of $\alpha$s and $\beta$s with specific particle labels,

$$\Theta = \alpha(1) \cdots \alpha(n/2 + S)\beta(n/2 + S + 1) \cdots \beta(n), \tag{5.92}$$

$$S_z \Theta = M_S \Theta, \tag{5.93}$$

$$M_S = S. \tag{5.94}$$

If now we operate upon $\Theta$ with $\mathcal{N}$ (corresponding to $\{n/2 + S, n/2 - S\}$) we obtain a function with $n/2 - S$ antisymmetric products of the $[\alpha\beta - \beta\alpha]$ sort,

$$\mathcal{N}\Theta = [\alpha(1)\beta(n/2 + S + 1) - \alpha(n/2 + S + 1)\beta(1)] \cdots \alpha(n/2 + S). \tag{5.95}$$

The spin raising operator (see Eq. (4.3)) may now be applied to this result, and we obtain

$$S^+ \mathcal{N}\Theta = \mathcal{N}S^+\Theta,$$
$$= 0, \qquad (5.96)$$

where we have used the commutation of $S^+$ with all permutations. The following argument indicates why the zero results. The terms of $S^+$ give zero with each $\alpha$ encountered but turn each $\beta$ encountered into an $\alpha$. Thus $S^+\Theta$ is $n/2 - S$ terms of products, each one of which has no more than $n/2 - S - 1$ $\beta$ functions in it. Considering how these would fit into the tableau shape, we see that there would have to be, for each term, one column in the tableau that has an $\alpha$ in both rows. This column, with its corresponding factor from $\mathcal{N}$, would thus appear as

$$[I - (ij)]\alpha(i)\alpha(j),$$

which is clearly zero. Eq. (5.96) is the consequence.

Thus, $\mathcal{N}\Theta$ is an eigenfunction of $\vec{S}^2$ because of Eq. (4.5),

$$\vec{S}^2 \mathcal{N}\Theta = S(S + 1)\mathcal{N}\Theta, \qquad (5.97)$$

and has total spin quantum number $S$ (also the $M_S$ value for this function). Other values of $M_S$ are available with $S^-$ should they be needed.

We now investigate the behavior of $\Theta$ when we apply our two simple Hermitian idempotents discussed earlier,

$$\Theta_{PNP} = \theta\mathcal{P}\mathcal{N}\mathcal{P}\Theta, \qquad (5.98)$$
$$= g_P\theta\mathcal{P}\mathcal{N}\Theta, \qquad (5.99)$$
$$\Theta_{NPN} = \theta'\mathcal{N}\mathcal{P}\mathcal{N}\Theta. \qquad (5.100)$$

Since $S^+$ and $S_z$ both commute with $\mathcal{N}$ and $\mathcal{P}$, both $\Theta_{PNP}$ and $\Theta_{NPN}$ are eigenfunctions of the $\vec{S}^2$ operator with total spin $S$ and $M_S = S$.

Heretofore in this section we have been working with the partition $\lambda = \{n/2 + S, n/2 - S\}$, but references to it in the equations have been suppressed. We now write $\Theta_{PNP}^\lambda$ and $\Theta_{NPN}^\lambda$. Applying the antisymmetrizer to the function of both space and spin that contains $\Theta_{PNP}^\lambda$,

$$\mathcal{A}\Psi_{PNP}^\lambda = \sum_{\lambda'ij} \frac{1}{f_{\lambda'}} e_{ij,r}^{\tilde{\lambda}'} \Xi e_{ij,s}^{\lambda'} \Theta_{PNP}^\lambda. \qquad (5.101)$$

If the antisymmetrizer has been conditioned (see Eqs. (5.75)–(5.78)) so that $e_{11,s}^{\lambda'}$ is $\theta\mathcal{P}\mathcal{N}\mathcal{P}^{\lambda'}$, we obtain

$$e_{ij,s}^{\lambda'}\Theta_{PNP}^\lambda = \delta_{1j}\delta_{\lambda\lambda'}e_{i1,s}^\lambda\Theta_{PNP}^\lambda, \qquad (5.102)$$

because of the orthogonality of the $e_{ij}^\lambda$ for different $\lambda$s.

We make a small digression and note that the *spin-degeneracy problem* we have alluded to before is evident in Eq. (5.102). It will be observed that $i = 1, \ldots, f_\lambda$ in the index of $e_{i1,s}^\lambda \Theta_{PNP}^\lambda$, and these functions are linearly independent since the $e_{ij,s}^\lambda$ are. There are, thus, $f_\lambda$ linearly independent spin eigenfunctions of eigenvalue $S(S + 1)$. Each of these has a full complement of $M_S$ values, of course. In view of Eq. (5.40) the number of spin functions increases rapidly with the number of electrons. Ultimately, however, the dynamics of a system governs if many or few of these are important.

Returning to our antisymmetrized function, we see it is now

$$\mathcal{A}\Psi_{PNP} = \frac{1}{f_\lambda} \sum_i e_{i1,r}^{\tilde{\lambda}} \Xi e_{i1,s}^\lambda \Theta_{PNP}^\lambda, \tag{5.103}$$

and we are in a position to examine its properties with regard to the Rayleigh quotient.

Considering first the denominator, we have

$$\langle \mathcal{A}\Psi_{PNP} | \mathcal{A}\Psi_{PNP} \rangle = f_\lambda^{-2} \sum_{ij} \{ \langle e_{i1,r}^{\tilde{\lambda}} \Xi | e_{j1,r}^{\tilde{\lambda}} \Xi \rangle$$

$$\times \langle e_{i1,s}^\lambda \Theta_{PNP}^\lambda | e_{j1,s}^\lambda \Theta_{PNP}^\lambda \rangle \}, \tag{5.104}$$

$$= f_\lambda^{-1} \langle \Xi | e_{11,r}^{\tilde{\lambda}} \Xi \rangle \langle \Theta_{PNP} | e_{11,s}^\lambda \Theta_{PNP}^\lambda \rangle, \tag{5.105}$$

since

$$\langle e_{i1,r}^{\tilde{\lambda}} \Xi | e_{j1,r}^{\tilde{\lambda}} \Xi \rangle = \langle \Xi | e_{1i,r}^{\tilde{\lambda}} e_{j1,r}^{\tilde{\lambda}} \Xi \rangle, \tag{5.106}$$

$$= \delta_{ij} \langle \Xi | e_{11,r}^{\tilde{\lambda}} \Xi \rangle, \tag{5.107}$$

with a very similar expression for the spin integral. Since the Hamiltonian of the ESE commutes with all permutations and symmetric group algebra elements, the same reductions apply to the numerator, and we obtain

$$\langle \mathcal{A}\Psi_{PNP} | H | \mathcal{A}\Psi_{PNP} \rangle = f_\lambda^{-1} \langle \Xi | H | e_{11,r}^{\tilde{\lambda}} \Xi \rangle \langle \Theta_{PNP}^\lambda | e_{11,s}^\lambda \Theta_{PNP}^\lambda \rangle. \tag{5.108}$$

This result should be carefully compared to that of Eq. (4.37), where there were two functions that have the same integral. Here we have $f_\lambda$ of them.[7]

Our final expression for the Rayleigh quotient is

$$E = \frac{\langle \mathcal{A}\Psi_{PNP} | H | \mathcal{A}\Psi_{PNP} \rangle}{\langle \mathcal{A}\Psi_{PNP} | \mathcal{A}\Psi_{PNP} \rangle}, \tag{5.109}$$

$$= \frac{\langle \Xi | H | e_{11}^{\tilde{\lambda}} \Xi \rangle}{\langle \Xi | e_{11}^{\tilde{\lambda}} \Xi \rangle}. \tag{5.110}$$

---

[7] We may note in passing that the partition for three electrons in a doublet state is {2,1} and $f_\lambda$ for this is 2. That is why we found two functions in our work in Chapter 4.

We are now done with spin functions. They have done their job to select the correct irreducible representation to use for the spatial part of the wave function. Since we no longer need spin, it is safe to suppress the $_s$ subscript in Eq. (5.110) and all of the succeeding work. We also note that the partition of the spatial function $\tilde{\lambda}$ is conjugate to the spin partition, i.e., $\{2^{n/2-S}, 2^{2S}\}$. From now on, if we have occasion to refer to this partition in general by symbol, we will drop the *tilde* and represent it with a bare $\lambda$.

### 5.5.2 The $\Xi$ function

We have so far said little about the nature of the space function, $\Xi$. Earlier we implied that it might be an orbital product, but this was not really necessary in our general work analyzing the effects of the antisymmetrizer and the spin eigenfunction. We shall now be specific and assume that $\Xi$ is a product of orbitals. There are many ways that a product of orbitals could be arranged, and, indeed, there are many of these for which the application of the $e_{11}^{\lambda}$ would produce zero. The partition corresponding to the spin eigenfunction had at most two rows, and we have seen that the appropriate ones for the spatial functions have at most two columns. Let us illustrate these considerations with a system of five electrons in a doublet state, and assume that we have five different (linearly independent) orbitals, which we label $a, b, c, d$, and $e$. We can draw two tableaux, one with the particle labels and one with the orbital labels,

$$\begin{bmatrix} a & b \\ c & d \\ e & \end{bmatrix} \quad \text{and} \quad \begin{bmatrix} 1 & 4 \\ 2 & 5 \\ 3 & \end{bmatrix}.$$

Associating symbols in corresponding positions from these two tableaux we may write down a particular product $\Xi = a(1)c(2)e(3)b(4)d(5)$. There are, of course, $5! = 120$ different arrangements of the orbitals among the particles, and all of the products are linearly independent. When we operate on them with the idempotent[8] $e_{11}$, however, the linear independence is greatly reduced and instead of 120 there are only $f = 5$ remaining.[9] This reduction is discussed in general by Littlewood[37]. For our work, however, we note that $e_{11} = m_{11} = u$, and $u\pi_i$ are linearly independent algebra elements. Therefore, using Eq. (5.64), the set consisting of the functions, $u\pi_i a(1)c(2)e(3)b(4)d(5); \; i = 1, \ldots, 5$ is linearly independent.[10] There are many sets of five that have this property, but we only need a set that spans the vector

---

[8] We now suppress the $\lambda$ superscript.
[9] At the beginning of Section 5.4.4 we saw that there were five standard tableau for the conjugate of the current shape.
[10] The linear independence of this sort of set is discussed in Section 5.4.5.

space of these functions, and the ones given here, based upon standard tableaux, will serve. We saw in Eq. (4.71) that there are two linearly independent orbital functions for the three-electron doublet state in the most general case, this being a consequence of the spin degeneracy of two. The result here is merely an extension. For the partition and tableau above the spin degeneracy is five, and the number of independent orbital functions is the same.

We saw in Chapter 4 that the number of independent functions is reduced to one if two of the three electrons are in the same orbital. A similar reduction occurs in general. In our five-electron example, if $b$ is set equal to $a$ and $c \neq d$, there are only two linearly independent functions, illustrating a specific case of the general result that the number of linearly independent functions arising from any orbital product is determined only by the orbitals "outside" the doubly occupied set. This is an important point, for which now we take up the general rules.

### 5.5.3 The independent functions from an orbital product

Assume we have a set of $m$ linearly independent orbitals. In order to do a calculation we must have $m \geq n/2 + S$, where $n$ is the number of electrons. Any fewer than this would require at least some *triple occupancy* of some of the orbitals, and any such product, $\Xi$, would yield zero when operated on by $u\pi_i$. This is the minimal number; ordinarily there will be more. Any particular product can be characterized by an *occupation vector*, $\vec{\gamma} = [\gamma_1 \, \gamma_2 \ldots \gamma_m]$ where $\gamma_i = 0, 1,$ or, $2$, and

$$\sum_{i=1}^{m} \gamma_i = n. \tag{5.111}$$

Clearly, the number of "2"s among the $\gamma_i$ cannot be greater than $n/2 - S$.

It is not difficult to convince oneself that functions with different $\vec{\gamma}$s are linearly independent. Therefore, the only cases we have to check are those produced from one occupation vector. Littlewood[37] shows how this is done considering *standard tableaux with repeated elements*. We choose an ordering for the labels of the orbitals we are using, $a_1 < a_2 < \cdots < a_k$; $n/2 + S \leq k \leq n$ that is arbitrary other than a requirement that the $a_i$ with $\gamma_i = 2$ occur first in the ordering.[11] We now place these orbitals in a tableau shape with the rule that all symbols are *nondecreasing* to the right in the rows and *definitely increasing* downward in the columns. Considering our five-electron case again, assume we have four orbitals $a < b < c < d$ and $a$ is doubly occupied. The rules for standard tableaux with repeated elements then

---

[11] This ordering can be quite arbitrary and, in particular, need not be related to an orbital's position in a product with a different $\vec{\gamma}$.

produce

$$
\begin{bmatrix} a & a \\ b & c \\ d & \end{bmatrix} \quad \text{and} \quad \begin{bmatrix} a & a \\ b & d \\ c & \end{bmatrix}
$$

as the two possible arrangements for this case. Using only three orbitals, $a < b < c$, with $a$ and $b$ doubly occupied, we obtain only one such tableau,

$$
\begin{bmatrix} a & a \\ b & b \\ c & \end{bmatrix}.
$$

The general result states that the number of linearly independent functions from the set $u\pi_i \,\Xi(\vec{\gamma})$; $i = 1, \ldots, f$ is the number of standard tableaux with repeated elements that can be constructed from the labels in the $\Xi$ product. As a general principle, this is not so easy to prove as some of the demonstrations of linear independence we have given above. The interested reader might, however, examine the case of two-column tableaux with which we are concerned. Examining the nature of the $\pi_i$ for this class of tableau, it is easy to deduce the result using $\mathcal{NPN}$. This is all that is needed, of course. The number of linearly independent functions cannot depend upon the representation.

We now see that for each $\vec{\gamma}$ we have $f^{\vec{\gamma}}$ linearly independent functions, $u\pi_i^{\vec{\gamma}} \,\Xi(\vec{\gamma})$, where $\pi_i^{\vec{\gamma}}$; $i = 1, \ldots, f^{\vec{\gamma}}$ is some subset[12] of all of the $\pi_i$ appropriate for $\Xi(\vec{\gamma})$.

The method for putting together a CI wave function is now clear. After choosing the $\vec{\gamma}$s to be included, one obtains

$$
u\Phi = u \sum_{i\vec{\gamma}} C_{i\vec{\gamma}} \pi_i^{\vec{\gamma}} \,\Xi(\vec{\gamma}), \tag{5.112}
$$

where the $C_{i\vec{\gamma}}$ are the linear variation parameters to optimize the energy. $u\Phi$ is thus a function satisfying the antisymmetry and spin conditions we choose and suitable for use with the ESE. We recall that $u\Phi$ is all that is needed to determine the energies. Minimizing the energy given by the Rayleigh quotient

$$
E = \frac{\langle u\Phi|H|u\Phi\rangle}{\langle u\Phi|u\Phi\rangle} \tag{5.113}
$$

$$
= \frac{\langle \Phi|H|u\Phi\rangle}{\langle \Phi|u\Phi\rangle} \tag{5.114}
$$

leads to a conventional nonorthogonal CI.

---

[12] We see now why there were relatively few spin functions generated by operators from the symmetric groups. For the partition $\{n/2 + S, n/2 - S\}$ and an $M_S = S$, there is only one standard tableau with repeated elements for the ordering $\alpha < \beta$. Thus only the $\pi_i^{-1}\mathcal{NPN}\Theta$ are linearly independent. All expressions of the form $\mathcal{NPN}\pi_j\Theta$, $\pi_j \neq I$ are zero.

Now that we know the number of linearly independent $n$-particle functions for a particular $\vec{\gamma}$, we can ask for the total number of linearly independent $n$-particle functions that can be generated from $m$ orbitals. Weyl[38] gave a general expression for all partitions and we will only quote his result for our two-column tableaux. The total number of functions, $D(n, m, S)$, i.e., the size of a full CI calculation is

$$D(n, m, S) = \frac{2S + 1}{m + 1} \binom{m + 1}{n/2 + S + 1} \binom{m + 1}{n/2 - S}. \tag{5.115}$$

This is frequently called the Weyl dimension formula. For small $S$ and large $m$ and $n$, $D$ can grow prodigiously beyond the capability of any current computer.[13]

### 5.5.4 Two simple sorts of VB functions

We saw in Section 5.4.7 that there were two Hermitian idempotents of the algebra that were easily constructed for each partition. Using these alternatives gives us two different (but equivalent) specific forms for the spatial part of the wave function. Specifically, if we choose $u = \theta \mathcal{N} \mathcal{P} \mathcal{N}$, we obtain the standard tableau functions introduced by the author and his coworkers[39]. If, on the other hand, we take $u' = \theta' \mathcal{P} \mathcal{N} \mathcal{P}$, we obtain the traditional Heitler–London–Slater–Pauling (HLSP) VB functions as discussed by Matsen and his coworkers[40]. In actual practice the $\pi_i^{\vec{\gamma}}$ for this case are not usually chosen from among those giving standard tableaux, but rather to give the Rumer diagrams (see Section 5.5.5). We asserted above that the permutations giving standard tableaux were only one possible set yielding linearly independent elements of the group algebra. This is a case in point. For the two-column tableaux the Rumer permutations are an alternative set that can be used, and are traditionally associated with different bonding patterns in the molecule.

Of these two schemes, it appears that the standard tableaux functions have properties that allow more efficient evaluation. This is directly related to the occurrence of the $\mathcal{N}$ on the "outside" of $\theta \mathcal{N} \mathcal{P} \mathcal{N}$. Tableau functions have the most antisymmetry possible remaining after the spin eigenfunction is formed. The HLSP functions have the least. Thus the standard tableaux functions are closer to single determinants, with their many properties that provide for efficient manipulation.[14] Our discussion of evaluation methods will therefore be focused on them. Since the two sets are equivalent, methods for writing the HLSP functions in terms of the others allow us to compare results for weights (see Section 1.1) of bonding patterns where this

---

[13] These considerations are independent of the nature of the orbitals other than their required linear independence. Thus, $D$ is the size of the full Hamiltonian matrix in either a VB treatment or an orthogonal molecular orbital CI.

[14] One may compare this difference with Goddard's[41] discussion of what he termed the G1 and Gf methods.

88  5 Advanced methods for larger molecules

is desired. If only energies and other properties calculated from expectation values are needed, the standard tableaux functions are sufficient.

We note finally that if $f^{\vec{\gamma}} = 1$ for a particular product function the standard tableaux function and HLSP function are the same.

### 5.5.5 Transformations between standard tableaux and HLSP functions

Since the standard tableaux functions and the HLSP functions span the same vector space, a linear transformation between them is possible. Specifically, it would appear that the task is to determine the $a_{ij}$s in

$$\theta \mathcal{N} \mathcal{P} \mathcal{N} \pi_i = \sum_j a_{ij} \theta' \mathcal{P} \mathcal{N} \mathcal{P} \rho_j, \tag{5.116}$$

where the $\pi_i$ are the permutations interconverting standard tableaux, and $\rho_j$ similarly interconvert Rumer diagrams. It turns out, however, that Eq. (5.116) cannot be valid. The difficulty arises because on the left of the equal sign the left-most operator is $\mathcal{N}$, while on the right it is $\mathcal{P}$. To see that Eq. (5.116) leads to a contradiction multiply both sides by $\mathcal{N}$. After factoring out some constants, one obtains

$$\theta \mathcal{N} \mathcal{P} \mathcal{N} \pi_i = \sum_j a_{ij} \frac{1}{g_{\mathcal{N}} g_{\mathcal{P}}} \mathcal{N} \mathcal{P} \rho_j, \tag{5.117}$$

which has a right hand side demonstratively different from that of Eq. (5.116). The left hand sides are, however, the same, so the two together lead to a contradiction. We must modify Eq. (5.116) by eliminating one or the other of the offending factors. It does not matter which, in principle, but the calculations are simpler if we use instead

$$\theta \mathcal{N} \mathcal{P} \mathcal{N} \pi_i = \sum_j a_{ij} \frac{f}{g} \mathcal{N} \mathcal{P} \rho_j. \tag{5.118}$$

In order to see why this modified problem actually serves our purpose, we digress to discuss some results for non-Hermitian idempotents.

The perceptive reader may already have observed that the functions we use can take many forms. Consider the non-Hermitian idempotent $(f/g)\mathcal{P} \mathcal{N}$. Using the permutations interconverting standard tableaux, one finds that $(f/g)\mathcal{P} \mathcal{N} \pi_i \Xi$; $i = 1, \ldots, f$ is a set of linearly independent functions (if $\Xi$ has no double occupancy). Defining a linear variation function in terms of these,

$$\Phi = \frac{f}{g} \mathcal{P} \mathcal{N} \sum_i a_i \pi_i \Xi, \tag{5.119}$$

one obtains for the matrix system,

$$H_{ij} = K \times \langle \pi_i \Xi | H \mathcal{N} \mathcal{P} \mathcal{N} | \pi_j \Xi \rangle, \qquad (5.120)$$

$$S_{ij} = K \times \langle \pi_i \Xi | \mathcal{N} \mathcal{P} \mathcal{N} | \pi_j \Xi \rangle, \qquad (5.121)$$

which is easily seen to be the same system as that obtained from the Hermitian idempotent, $\theta \mathcal{N} \mathcal{P} \mathcal{N}$. The "$K$" is different, of course, but this cancels between the numerator and denominator of the Rayleigh quotient. Thus,

$$\Psi' = \frac{f}{g g_N} \mathcal{N} \mathcal{P} \mathcal{N} \sum_i a_i \pi_i \Xi \qquad (5.122)$$

will produce the same eigensystem and eigenvectors as the variation function $\Psi$ of Eq. (5.119), but the resulting spatial functions are not equal, $\Psi \neq \Psi'$. Some considerable care is required in interpreting this result. It must be remembered that the spatial functions under discussion are only a fragment of the total wave function, and are related to expectation values of the total wave function only if the operator involved commutes with all permutations of $S_n$. There are two important cases that demonstrate the care that must be used in this matter.

Consider an operator commonly used to determine the charge density:

$$D_{op} = \sum_i \delta(\vec{r}_i - \vec{\rho}), \qquad (5.123)$$

where $\vec{\rho}$ is the position at which the density is given and $i$ now labels electrons. This operator commutes with all permutations and is thus satisfactory for determining the charge density from $\Psi$, $\Psi'$, or the whole wave function. The spatial probability density is another matter. In this case the operator is

$$P_{op} = \prod_i \delta(\vec{r}_i - \vec{\rho}_i), \qquad (5.124)$$

where the $\vec{\rho}_i$ are the values at which the functions are evaluated. As it stands, this is satisfactory for the whole wave function, but for neither $\Psi$ nor $\Psi'$. To work with the latter two, we must make it commute with all permutations, and it must be modified to

$$P'_{op} = \frac{1}{n!} \sum_{\tau \in S_n} \tau^{-1} \prod_i \delta(\vec{r}_i - \vec{\rho}_i) \tau, \qquad (5.125)$$

where the permutations do not operate on the $\vec{\rho}_i$. The $P'_{op}$ form gives the same value in all three cases.

After this digression we now return to the problem of determining the HLSP functions in terms of the standard tableaux functions. We solve Eq. (5.118) by multiplying both sides by $\pi_k^{-1}$ and evaluating both sides for the identity element:[15]

$$\frac{1}{g_N}[\![\mathcal{NPN}\pi_i\pi_k^{-1}]\!] = \sum_j [\![\mathcal{NP}\rho_j\pi_k^{-1}]\!]a_{ji} \qquad (5.126)$$

$$M_{ki} = \sum_j B_{kj}a_{ji}, \qquad (5.127)$$

and denoting by $A$ the matrix with elements $a_{ij}$, we obtain

$$A = B^{-1}M. \qquad (5.128)$$

In Eq. (5.128)

$$M_{ki} = \frac{1}{g_N}[\![\pi_k^{-1}\mathcal{NPN}\pi_i]\!], \qquad (5.129)$$

is the "overlap" matrix for $\theta\mathcal{NPN}$ (see Eq. (5.73) and following).

For singlet systems the bonding patterns for Rumer diagrams are conventionally obtained by writing the symbols for the orbitals in a ring (shown here for six), and drawing all diagrams where all pairs of orbital symbols are connected by a line and no lines cross[2, 13].

Our treatment has been oriented towards using tableaux to represent functions rather than Rumer diagrams, and it will be convenient to continue. Thus, corresponding to the five canonical diagrams for a ring of six orbital symbols one can write

$$\begin{bmatrix} a & b \\ c & d \\ e & f \end{bmatrix}_R \quad \begin{bmatrix} a & f \\ b & c \\ d & e \end{bmatrix}_R \quad \begin{bmatrix} a & b \\ c & f \\ d & e \end{bmatrix}_R \quad \begin{bmatrix} a & d \\ b & c \\ e & f \end{bmatrix}_R \quad \begin{bmatrix} a & f \\ b & f \\ c & d \end{bmatrix}_R$$

where the symbols in the same row are "bonded" in the Rumer diagram. We have made a practice in using [ ] around our tableaux, and those that refer to functions where we use $\mathcal{PNP}$ will be given "$R$" subscripts to distinguish them from functions where we have used $\mathcal{NPN}$. This notational device will be used extensively in Part II of the book where many comparisons between standard tableaux functions and HLSP functions are shown.

---

[15] We commented above that the form of Eq. (5.118) was simpler than the result of removing $\mathcal{N}$ from the other side. This arises because determining $[\![\mathcal{PNP}\tau]\!]$ is, in general, much more difficult than evaluating $[\![\mathcal{NP}\tau]\!]$, because simple expressions for $\mathcal{PNP}$ are known only for singlet and doublet systems.

The tableaux in the last paragraph are, of course, not unique. In any row either orbital could be written first, and any order of rows is possible. Thus, there are $2^3 \times 3! = 48$ different possible arrangements for each. We have made them unique by setting $a < b < c < d < e < f$, making each row increase to the right and the first column increase downward. These are not standard tableaux – the second column is not alway increasing downward. Using

$$\begin{bmatrix} 1 & 4 \\ 2 & 5 \\ 3 & 6 \end{bmatrix}$$

for the particle label tableau, it is seen that the permutations $I$, $(25\,364)$, $(365)$, $(254)$, and $(23\,564)$ will generate all five orbital tableaux from the first, and can be used for the $\rho_i$ of Eq. (5.118).

This transformation is tedious to obtain by hand, and computer programs are to be preferred. A few special cases have been given[39]. An example is also given in Section 6.3.2.

### 5.5.6 Representing $\theta \, \mathcal{N} \mathcal{P} \mathcal{N} \, \Xi$ as a functional determinant

For the efficient evaluation of matrix elements, it is useful to have a representation of $\theta \mathcal{N} \mathcal{P} \mathcal{N} \Xi$ as a functional determinant. We consider subgroups and their cosets to obtain the desired form.

The operator $\mathcal{N}$ consists of terms for all of the permutations of the subgroup $G_\mathcal{N}$, and $\mathcal{P}$ those for the subgroup $G_\mathcal{P}$. Except for the highest multiplicity case, $S = n/2$, $G_\mathcal{N}$ is smaller than the whole of $S_n$. Let $\rho_\mathcal{N} \in G_\mathcal{N}$ and $\tau_1 \notin G_\mathcal{N}$. Consider all of the permutations $\rho_\mathcal{N} \tau_1$ for fixed $\tau_1$ as $\rho_\mathcal{N}$ runs over $G_\mathcal{N}$. This set of permutations is called a *right coset* of $G_\mathcal{N}$. The designation as "right" arises because $\tau_1$ is written to the right of all of the elements of $G_\mathcal{N}$. We abbreviate the right coset as $G_\mathcal{N} \tau_1$. There is also a *left coset* $\tau_1 G_\mathcal{N}$, not necessarily the same as the right coset. Consider a possibly different right coset $G_\mathcal{N} \tau_2$, $\tau_2 \notin G_\mathcal{N}$. This set is either completely distinct from $G_\mathcal{N} \tau_1$ or identical with it. Thus, assume there is one permutation in common between the two cosets,

$$\rho_1 \tau_1 = \rho_2 \tau_2; \ \rho_1, \rho_2 \in G_\mathcal{N} \tag{5.130}$$

$$\rho_3 \rho_1 \tau_1 = \rho_3 \rho_2 \tau_2; \ \rho_3 \in G_\mathcal{N}, \tag{5.131}$$

and, as $\rho_3$ ranges over $G_\mathcal{N}$, the right and left hand sides of Eq. (5.131) run over the two cosets and we see they are the same except possibly for order. The test may be stated another way: if

$$\tau_2 \tau_1^{-1} = \rho_2^{-1} \rho_1 \in G_\mathcal{N}, \tag{5.132}$$

$\tau_1$ and $\tau_2$ generate the same coset.

We conclude that one can find a number of right coset generators giving distinct cosets until the permutations of $S_n$ are exhausted. Symbolizing the right coset generators as $\tau_1 = I, \tau_2, \ldots, \tau_p$, we have

$$S_n = G_\mathcal{N} I \oplus G_\mathcal{N} \tau_2 \oplus \cdots \oplus G_\mathcal{N} \tau_p, \tag{5.133}$$

where the first coset is $G_\mathcal{N}$ itself. This leads to the often quoted result that the order of any subgroup must be an integer divisor of the order of the whole group and, in this case, we have

$$p = \binom{n}{n/2 - S}. \tag{5.134}$$

Our goal now is to find a convenient set of right coset generators for $G_\mathcal{N}$ that gives $S_n$. Let us now consider specifically the case for the $\{2^k \, 1^{n-2k}\}$ partition with $k = n/2 - S$, and the tableau,

$$
\begin{bmatrix}
1 & n-k+1 \\
\vdots & \vdots \\
k & n \\
k+1 & \\
\vdots & \\
n-k &
\end{bmatrix}.
$$

The order of $G_\mathcal{N}$ is $g_\mathcal{N} = (n-k)!k!$. Now let $i_1, i_2, \ldots, i_l$ be $l \leq k$ of the integers from the first column of our tableau and let $j_1, j_2, \ldots, j_l$ be the same number from the second column. These two sets of integers define a special permutation $[(i)(j)]_l = (i_1 j_1) \cdots (i_l j_l)$, which is a product of $l$ noninteracting binaries. Since each binary contains a number from each column, none with $l > 0$ are members of $G_\mathcal{N}$. Some, but not all, are members of $G_P$, however. Amongst all of these there is a subset that we call *canonical* in which $i_1 < i_2 < \cdots < i_l$ and $j_1 < j_2 < \cdots < j_l$. The number of these is

$$\binom{n-k}{l}\binom{k}{l},$$

and it is easily shown that

$$\sum_{l=0}^{k} \binom{n-k}{l}\binom{k}{l} = \binom{n}{k}. \tag{5.135}$$

Thus, if the distinct canonical $[(i)(j)]$ generate distinct cosets, we have all of them, since

$$g_N \begin{pmatrix} n \\ k \end{pmatrix} = n!. \tag{5.136}$$

Considering $[(i)(j)]_l$ and $[(i')(j')]_{l'}$, we use Eq. (5.132) to test whether they generate the same coset. The $[(i)(j)]_l$ are, of course, their own inverses, and for the present test we have

$$[(i)(j)]_l[(i')(j')]_{l'} \overset{?}{\in} G_N. \tag{5.137}$$

If $[(i)(j)]_l$ and $[(i')(j')]_{l'}$ have any binaries with no numbers in common then these will remain unaffected in the product, and since none of the binaries is a member of $G_N$, neither is the product and the cosets must be different. If there are any binaries in common these cancel and there remain only binaries that have numbers connected in one or more chains. Consider a simple two-member chain, $(a\ b)(c\ b) = (a\ c)(a\ b)$. The binary $(a\ c) \in G_N$ but $(a\ b)$ is not, so this chain cannot be a member of $G_N$, and, again, the cosets are different. Our simple two-member chain could, however, be the start of a longer one, and proceeding this way we see that we always arrive at the conclusion that the canonical $[(i)(j)]_l$s generate all of the cosets.

Going back to $\theta N P N$, we write out the $N$ on the right explicitly and carry out a number of transformations.

$$\theta N P N = \theta N P \sum_{v \in G_N} (-1)^{\sigma_v} v, \tag{5.138}$$

$$= \theta N \sum_{v \in G_N} (-1)^{\sigma_v} v v^{-1} P v, \tag{5.139}$$

$$= \frac{f}{n!} N B, \tag{5.140}$$

$$B = \frac{1}{g_N} \sum_{v \in G_N} v^{-1} P v, \tag{5.141}$$

and we see that the $B$ operator is a sort of symmetrization of the $P$ operator. We note first that $[\![B]\!] = 1$. The operator $P$ is a sum of terms

$$P = I + \sum_{l=1}^{k} p_l, \tag{5.142}$$

where $p_l$ is a sum over all of the sort of $[(i)(j)]_l$ that correspond to $l$ and have "horizontal" binaries only. There are $\binom{k}{l}$ permutations in $p_l$. Next, we observe that each term in Eq. (5.141) has the form

$$\sum_{v} v^{-1} p_l v = (n-k)!k! \binom{k}{l} \binom{n-k}{l}^{-1} \binom{k}{l}^{-1} b_l, \qquad (5.143)$$

$$= g_N \binom{n-k}{l}^{-1} b_l, \qquad (5.144)$$

where $b_l$ is the sum of all of the coset generators corresponding to $l$. Equation (5.143) is obtained merely by the correct counting: the factors on the right are the number of terms in the sum and $p_l$ divided by the number of terms in $b_l$. Thus,

$$\theta \mathcal{N} \mathcal{P} \mathcal{N} = \frac{f}{g} \mathcal{N} \mathcal{B}, \qquad (5.145)$$

$$= \frac{f}{g} \mathcal{N} \sum_{l=0}^{k} \binom{n-k}{l}^{-1} b_l, \qquad (5.146)$$

$$= \frac{f}{g} \mathcal{B} \mathcal{N}, \qquad (5.147)$$

where we know $\mathcal{N}$ and $\mathcal{B}$ commute, since they are both Hermitian and so is $\mathcal{N} \mathcal{P} \mathcal{N}$.

As an example of how $\mathcal{N}$ and $b_l$ operators work together we observe that the full antisymmetrizer corresponding to $S_n$ may be written with $\mathcal{N}$ and the $b_l$ operators,

$$\mathcal{A} = \frac{1}{n!} \mathcal{N} \sum_{l=0}^{k} (-1)^l b_l, \qquad (5.148)$$

since the right hand side has each permutation once and each will have the correct sign. We emphasize that this is valid for any $k$.

Now consider $n$ functions $u_1, u_2, \ldots, u_n$ and form the $n$-particle product function $\Xi = u_1(1)u_2(2)\cdots u_n(n)$. Using the form of the antisymmetrizer of Eq. (5.148) we see that

$$\mathcal{A}\Xi = \frac{1}{n!} \begin{vmatrix} u_1(1) & \cdots & u_n(1) \\ \vdots & & \vdots \\ u_1(n) & \cdots & u_n(n) \end{vmatrix}, \qquad (5.149)$$

and for each $k$ of Eq. (5.148) we have a way of representing a determinant. These correspond to different Lagrange expansions that can be used to evaluate determinants, and, in particular, the use of $k = 1$ is closely associated with Cramer's rule[42].

We now define another operator (group algebra element) using the $b_l$ coset generator sums,

$$\mathcal{D}(q) = \sum_{l=0}^{k} q^l b_l, \qquad (5.150)$$

where $q$ could be complex. With $\mathcal{N}$ this new operator may be applied to the orbital product $\Xi$. A little reflection will convince the reader that the result may be written as a functional determinant,

$$\mathcal{N}\mathcal{D}(q)\Xi = \begin{vmatrix} P & Q \\ -qR & S \end{vmatrix}, \tag{5.151}$$

where $P$, $Q$, $R$, and $S$ are blocks from the determinant in Eq. (5.149). Their sizes and shapes depend upon $k$: $P$ is $(n-k) \times (n-k)$, $Q$ is $(n-k) \times k$, $R$ is $k \times (n-k)$, and $S$ is $k \times k$. The block $-qR$ represents the variable $-q$ multiplying each function in the $R$-block. We note that if $q = -1$ the operator $\mathcal{D}(q)$ is just the sum of coset generators in Eq. (5.148), and the determinant in Eq. (5.151) just becomes the one in Eq. (5.149).

We may now use the $\beta$-function integral[28],

$$\int_0^1 t^l (1-t)^{n-k-l} dt = (n-k+1)^{-1} \binom{n-k}{l}^{-1}, \tag{5.152}$$

and, letting $q = t/(1-t)$, convert $\mathcal{D}(q)$ to $\mathcal{B}$. Thus, one obtains

$$(n-k+1) \int_0^1 (1-t)^{(n-k)} \mathcal{D}[(t/(1-t))] dt = \mathcal{B}. \tag{5.153}$$

Putting together these results, we obtain the expression for $\theta\mathcal{N}\mathcal{P}\mathcal{N}\Xi$ as the integral of a functional determinant,

$$\theta\mathcal{N}\mathcal{P}\mathcal{N}\Xi = \frac{(n-k+1)f}{g} \int_0^1 (1-t)^{(n-k)} \begin{vmatrix} P & Q \\ -qR & S \end{vmatrix} dt, \tag{5.154}$$

$$q = \frac{t}{1-t}. \tag{5.155}$$

The same sort of considerations allow one to determine matrix elements. Let $v_1(1) \cdots v_n(n) = \Upsilon$ be another orbital product. There is a joint overlap matrix between the $v$- and $u$-functions:

$$S(\bar{v}, \bar{u}) = \begin{bmatrix} \langle v_1|u_1\rangle & \cdots & \langle v_1|u_n\rangle \\ \vdots & & \vdots \\ \langle v_n|u_1\rangle & \cdots & \langle v_n|u_n\rangle \end{bmatrix}, \tag{5.156}$$

and we may use it to assemble a functional determinant. Thus, we have

$$\langle \Upsilon|\theta\mathcal{N}\mathcal{P}\mathcal{N}\Xi\rangle = \frac{(n-k+1)f}{g} \int_0^1 (1-t)^{(n-k)} \begin{vmatrix} P' & Q' \\ -qR' & S' \end{vmatrix} dt, \tag{5.157}$$

$$q = \frac{t}{1-t}, \tag{5.158}$$

where the primed blocks of the determinant come from $S(\bar{v}, \bar{u})$ in the same way as the blocks of Eq. (5.151) were obtained from the determinant of Eq. (5.149). Somewhat more complicated but similar considerations provide for the determination of the matrix elements of the Hamiltonian.

When there are doubly occupied orbitals among the $u_i$, simplifications occur in these expressions. In addition, the integrand is a polynomial in $t$ of degree $n - k$, and may be evaluated exactly in computer applications using an optimal Gauss quadrature formula[28]. For further details on both these points, the reader is referred to the literature[39]. The VB calculations reported on in Part II of the book were all carried out by a computer program implementing the discussion of this section.

# 6

# Spatial symmetry

Spatial symmetry plays a role in a large number of the examples in Part II of this book. This can arise in a number ways, but the two most important are simplification of the calculations and labeling of the energy states. We have devoted considerable time and space in Chapter 5 to the effects of identical particle symmetry and spin. In this chapter we look at some of the ways spatial symmetry interacts with anti-symmetrization.

We first note that spatial symmetry operators and permutations commute when applied to the functions we are interested in. Consider a multiparticle function $\phi(\vec{r}_1, \vec{r}_2, \dots, \vec{r}_n)$, where each of the particle coordinates is a 3-vector. Applying a permutation to $\phi$ gives

$$\pi\phi(\vec{r}_1, \vec{r}_2, \dots, \vec{r}_n) = \phi(\vec{r}_{\pi_1}, \vec{r}_{\pi_2}, \dots, \vec{r}_{\pi_n}), \tag{6.1}$$

where $\{\pi_1, \pi_2, \dots, \pi_n\}$ is some permutation of the set $\{1, 2, \dots, n\}$. Now consider the result of applying a spatial symmetry operator,[1] i.e., a rotation, reflection, or rotary-reflection, to $\phi$. Symbolically, we write for a spatial symmetry operation, $R$,

$$R\vec{r} = \vec{r'}, \tag{6.2}$$

$$R\phi(\vec{r}_1, \vec{r}_2, \dots, \vec{r}_n) = \phi(\vec{r'}_1, \vec{r'}_2, \dots, \vec{r'}_n), \tag{6.3}$$

and we see that

$$\pi R\vec{r} = R\pi\vec{r}, \tag{6.4}$$

$$= \phi(\vec{r'}_{\pi_1}, \vec{r'}_{\pi_2}, \dots, \vec{r'}_{\pi_n}). \tag{6.5}$$

---

[1] In physics and chemistry there are two different forms of spatial symmetry operators: the direct and the indirect. In the direct transformation, a rotation by $\pi/3$ radians, e.g., causes all vectors to be rotated around the rotation axis by this angle with respect to the coordinate axes. The indirect transformation, on the other hand, involves rotating the coordinate axes to arrive at new components for the same vector in a new coordinate system. The latter procedure is not appropriate in dealing with the electronic factors of Born–Oppenheimer wave functions, since we do not want to have to express the nuclear positions in a new coordinate system for each operation.

## 6.1 The AO basis

The functions we use are products of AOs, and, to be useful in a calculation, the AOs must be a basis for a representation of the spatial group. Since the spatial operations and permutations commute, the tableau functions we use also provide a basis for a representation of the spatial group. This is generally true regardless of the nature of the representation provided by the AOs themselves[43]. Nevertheless, to work with tableaux on computers it greatly simplifies programs if the AO basis provides a representation of a somewhat special sort we call *generalized permutation*. If we have an appropriate AO basis, it supports a unitary representation of the spatial group $G_S = \{I, R_2, R_3, \ldots, R_f\}$,

$$R_i \chi_j = \sum_k \chi_k D(R_i)_{kj}, \tag{6.6}$$

where $\chi_j$ are the AOs and the $D(R_i)_{kj}$ are, in general, reducible. $D(R_i)$ is a generalized permutation matrix if every element is either zero or a number of unit magnitude. Because of the unitarity, each row or column of $D(R_i)$ has exactly one nonzero element, and this one is $\pm 1$. As it turns out, this is not an extremely special requirement, but it is not always possible to arrange. The following are some guidelines as to when it *is* possible.[2]

- $G_S$ is abelian.
- $G_S$ has a principal rotation axis of order $>2$, and no atoms of the molecule are centered on it. This frequently requires the coordinate axes for the AOs to be different on different atoms.
- $G_S$ transforms the $x$-, $y$-, and $z$-coordinate axes into $\pm$ themselves, and we use tensorial rather than spherical $d, f, \ldots$ functions. That is, our $d$-set transforms as $\{x^2, y^2, z^2, xy, xz, yz\}$ with similar sets for the higher $l$-values.

In cases where these guidelines cannot be met, one must use the largest abelian subgroup from the true $G_S$ of the molecule. We will show some examples later.

## 6.2 Bases for spatial group algebras

Just as we saw with the symmetric groups, groups of spatial operations have associated group algebras with a matrix basis for this algebra,

$$e_{ij}^\alpha = \frac{f_\alpha}{g} \sum_{i=1}^g D(R_i)_{ij}^{\alpha*} R_i. \tag{6.7}$$

---

[2] We emphasize these rules are not needed theoretically. They are merely those that the symmetry analysis in CRUNCH requires to work.

This should be compared with Eq. (5.19), but in this case we can assume that the irreducible representation is unitary without causing any complications. The law of combination is identical with the earlier Eq. (5.20),

$$e_{ij}^{\alpha} e_{kl}^{\beta} = \delta_{jk} \delta_{\alpha\beta} e_{il}^{\alpha}. \tag{6.8}$$

We use the same symbol for the two kinds of groups. This normally causes no confusion. These operators of course satisfy

$$\left(e_{ij}^{\alpha}\right)^{\dagger} = e_{ji}^{\alpha}, \tag{6.9}$$

and, thus, $e_{ii}^{\alpha}$ is Hermitian. All of the $e_{ij}^{\alpha}$ also commute with the Hamiltonian.

The element $e_{11}^{\alpha}$ is a projector for the first component of the $\alpha^{\text{th}}$ irreducible representation basis. Using standard tableaux functions we can select a function of a given symmetry *and* a given spin state with

$$\psi_j^{\alpha} = e_{11}^{\alpha} \theta \mathcal{N} \mathcal{P} \mathcal{N} T_j, \tag{6.10}$$

where $T_j$ is a product of AOs associated with the $j^{\text{th}}$ standard tableau. When we evaluate matrix elements of either the overlap or the Hamiltonian between two functions of these types we have

$$\langle \psi_j^{\alpha} | \psi_k^{\beta} \rangle = \langle e_{11}^{\alpha} \theta \mathcal{N} \mathcal{P} \mathcal{N} T_j | e_{11}^{\beta} \theta \mathcal{N} \mathcal{P} \mathcal{N} T_k \rangle, \tag{6.11}$$

$$= \delta_{\alpha\beta} \langle T_j | e_{11}^{\beta} \theta \mathcal{N} \mathcal{P} \mathcal{N} T_k \rangle, \tag{6.12}$$

$$\langle \psi_j^{\alpha} | H | \psi_k^{\beta} \rangle = \delta_{\alpha\beta} \langle T_j | H | e_{11}^{\beta} \theta \mathcal{N} \mathcal{P} \mathcal{N} T_k \rangle. \tag{6.13}$$

## 6.3 Constellations and configurations

In quantum mechanical structure arguments we often speak of a configuration as a set of orbitals with a particular pattern of occupations. In this sense, if we consider the first of a set of standard tableaux, $T_1$, we can see that it establishes a configuration of orbitals. The other standard tableaux, $T_2, \ldots, T_f$, all establish the same configuration. Consider, however, the result of operating on $T_1$ with an element of $G_S$. It is simple to see why the assumption that the representation $D(R)$ in Eq. (6.6) consists of generalized permutation matrices simplifies the result of this operation: in this case $R_i T_1$ is just another product function $\pm T'$. It may involve the same configuration or a different one, but it is just a simple product function. We use the term *constellation* to denote the collection of configurations that are generated by all of the elements operating upon $R_i T_1$; $i = 1, 2, \ldots, g$. Putting this another way, a constellation is a set of configurations closed under the operations of $G_S$. It will be useful to illustrate some of these ideas with examples. We give

Table 6.1. *Transformation of $H_2O$ AOs.*

| $I$ | $C_2$ | $\sigma_{xz}$ | $\sigma_{zy}$ |
|-----|-------|---------------|---------------|
| $2s$    | $2s$     | $2s$     | $2s$     |
| $2p_x$  | $-2p_x$  | $2p_x$   | $-2p_x$  |
| $2p_y$  | $-2p_y$  | $-2p_y$  | $2p_y$   |
| $2p_z$  | $2p_z$   | $2p_z$   | $2p_z$   |
| $1s_a$  | $1s_b$   | $1s_b$   | $1s_a$   |
| $1s_b$  | $1s_a$   | $1s_a$   | $1s_b$   |

three: a $C_{2v}$ system, $H_2O$; a $C_{3v}$ system, $NH_3$; and a $D_{6h}$ system, the $\pi$ system of benzene.

### 6.3.1 Example 1. $H_2O$

Consider a water molecule with a minimal basis on the atoms. We have a $1s$, $2s, 2p_x, 2p_y, 2p_z$ set on the O atom and $1s_a$ and $1s_b$ on the H atoms. We assume the molecule is oriented in the $y$–$z$ plane with the O on the $z$-axis and the center of mass at the origin of a right-handed Cartesian coordinate system. It does not detract from this illustration if we ignore the O$1s$, and we suppress them from all tableaux. $H_2O$ belongs to the $C_{2v}$ symmetry group, which is abelian and, hence, satisfies one of our guidelines above. Table 6.1 gives the transformation of the AOs under the operations of the group.

Consider a configuration $2s^2 2p_x^2 1s_a^2 2p_y 2p_z$. The identity and $\sigma_{zy}$ operations leave it unchanged and the other two give $2s^2 2p_x^2 1s_b^2 2p_y 2p_z$, and these configurations comprise one of the constellations for $H_2O$ and this basis. The projector for the $A_1$ symmetry species of $C_{2v}$ is

$$e^{A_1} = \frac{1}{4}(I + C_2 + \sigma_{xz} + \sigma_{zy}), \qquad (6.14)$$

and taking

$$\begin{bmatrix} 2s & 2s \\ 2p_x & 2p_x \\ 1s_a & 1s_a \\ 2p_y & 2p_z \end{bmatrix}$$

as the defining tableau, we obtain

$$e^{A_1} \begin{bmatrix} 2s & 2s \\ 2p_x & 2p_x \\ 1s_a & 1s_a \\ 2p_y & 2p_z \end{bmatrix} = \frac{1}{2}\left( \begin{bmatrix} 2s & 2s \\ 2p_x & 2p_x \\ 1s_a & 1s_a \\ 2p_y & 2p_z \end{bmatrix} - \begin{bmatrix} 2s & 2s \\ 2p_x & 2p_x \\ 1s_b & 1s_b \\ 2p_y & 2p_z \end{bmatrix} \right), \qquad (6.15)$$

as the $A_1$ symmetry function based upon this constellation. If, alternatively, we used the $B_2$ projector,

$$e^{B_2} = \frac{1}{4}(I - C_2 - \sigma_{xz} + \sigma_{zy}),$$  (6.16)

we would obtain the same two tableaux as in Eq. (6.15), but with a $+$ sign between them. The other two projectors yield zero.

The symmetry standard tableaux functions are not always so intuitive as those in the first case we looked at. Consider, e.g., the configuration $2s2p_x^2 2p_y^2 2p_z 1s_a 1s_b$, for which there are two standard tableaux and no other members in the constellation,

$$\begin{bmatrix} 2p_x & 2p_x \\ 2p_y & 2p_y \\ 2s & 2p_z \\ 1s_a & 1s_b \end{bmatrix} \quad \text{and} \quad \begin{bmatrix} 2p_x & 2p_x \\ 2p_y & 2p_y \\ 2s & 1s_a \\ 2p_z & 1s_b \end{bmatrix}.$$

When we apply $e^{A_1}$ to the first of these, we obtain

$$e^{A_1} \begin{bmatrix} 2p_x & 2p_x \\ 2p_y & 2p_y \\ 2s & 2p_z \\ 1s_a & 1s_b \end{bmatrix} = \frac{1}{2} \left( \begin{bmatrix} 2p_x & 2p_x \\ 2p_y & 2p_y \\ 2s & 2p_z \\ 1s_a & 1s_b \end{bmatrix} + \begin{bmatrix} 2p_x & 2p_x \\ 2p_y & 2p_y \\ 2s & 2p_z \\ 1s_b & 1s_a \end{bmatrix} \right),$$  (6.17)

where the second term on the right is *not* a standard tableau, but may be written in terms of them. Using the methods of Chapter 5 we find that

$$\begin{bmatrix} 2p_x & 2p_x \\ 2p_y & 2p_y \\ 2s & 2p_z \\ 1s_b & 1s_a \end{bmatrix} = \begin{bmatrix} 2p_x & 2p_x \\ 2p_y & 2p_y \\ 2s & 2p_z \\ 1s_a & 1s_b \end{bmatrix} - \begin{bmatrix} 2p_x & 2p_x \\ 2p_y & 2p_y \\ 2s & 1s_a \\ 2p_z & 1s_b \end{bmatrix},$$  (6.18)

and thus

$$e^{A_1} \begin{bmatrix} 2p_x & 2p_x \\ 2p_y & 2p_y \\ 2s & 2p_z \\ 1s_a & 1s_b \end{bmatrix} = \begin{bmatrix} 2p_x & 2p_x \\ 2p_y & 2p_y \\ 2s & 2p_z \\ 1s_a & 1s_b \end{bmatrix} - \frac{1}{2} \begin{bmatrix} 2p_x & 2p_x \\ 2p_y & 2p_y \\ 2s & 1s_a \\ 2p_z & 1s_b \end{bmatrix},$$  (6.19)

which is a projected symmetry function, although not manifestly so.

It is not difficult to show that

$$e^{A_1} \begin{bmatrix} 2p_x & 2p_x \\ 2p_y & 2p_y \\ 2s & 1s_a \\ 2p_z & 1s_b \end{bmatrix} = 0,$$  (6.20)

and the second standard tableau does not contribute to $^1A_1$ wave functions. This result indicates that the first standard tableau is not by itself a pure symmetry

Table 6.2. *Transformation of NH$_3$ AOs.*

| $I$ | $C_3$ | $C_3^2$ | $\sigma_x$ | $\sigma_y{}^a$ | $\sigma_z{}^a$ |
|---|---|---|---|---|---|
| $2s$ | $2s$ | $2s$ | $2s$ | $2s$ | $2s$ |
| $2p_x$ | $2p_y$ | $2p_z$ | $2p_x$ | $2p_z$ | $2p_y$ |
| $2p_y$ | $2p_z$ | $2p_x$ | $2p_z$ | $2p_y$ | $2p_x$ |
| $2p_z$ | $2p_x$ | $2p_y$ | $2p_y$ | $2p_x$ | $2p_z$ |
| $1s_x{}^b$ | $1s_y$ | $1s_z$ | $1s_x$ | $1s_z$ | $1s_y$ |
| $1s_y{}^b$ | $1s_z$ | $1s_x$ | $1s_z$ | $1s_y$ | $1s_x$ |
| $1s_z{}^b$ | $1s_x$ | $1s_y$ | $1s_y$ | $1s_x$ | $1s_z$ |

$^a$ Each reflection plane is labeled with the coordinate
axis that is contained in it.
$^b$ Each H-atom orbital is labeled with the reflection
plane it resides on.

type but contains $A_1$ and $B_2$ components, while the second is pure $B_2$. The linear
combination of Eq. (6.19) removes the unwanted part from the first tableau.

We emphasize that these results are specific to the way we have ordered the
particle numbers in the AOs. Other arrangements could give results that look quite
different, but which would, nevertheless, be equivalent as far as giving the same
eigenvalues of the ESE is concerned.

### 6.3.2 Example 2. NH$_3$

$C_{3v}$ is not an abelian group, but it is not difficult to orient a minimal basis involving
$s$ and $p$ orbitals to make the representation of the AO basis a set of generalized
permutation matrices. We orient the $C_3$-axis of the group along the unit vector
$\{1/\sqrt{3}, 1/\sqrt{3}, 1/\sqrt{3}\}$. The center of mass is at the origin and the N atom is on the
$C_3$-axis in the negative direction from the origin. The three reflection planes of the
group may be defined by the rotation axis and the three coordinate axes, respectively.
There is an H atom in each of the reflection planes at an N—H bond distance from
the N atom and at an angle of $\approx 76°$ from the rotation axis. In our description we
suppress the closed $1s^2$ core as before. Table 6.2 shows the transformation properties
of the basis. We consider the configuration $2s^2 2p_x 2p_y 2p_z 1s_x 1s_y 1s_z$, which is the
only member of its constellation. Once we have chosen a specific arrangement for
the first tableau, the other four standard tableaux may be given

$$
\begin{bmatrix} 2s & 2s \\ 2p_x & 1s_x \\ 2p_y & 1s_y \\ 2p_z & 1s_z \end{bmatrix}
\begin{bmatrix} 2s & 2s \\ 2p_x & 1s_x \\ 2p_y & 2p_z \\ 1s_y & 1s_z \end{bmatrix}
\begin{bmatrix} 2s & 2s \\ 2p_x & 2p_y \\ 1s_x & 1s_y \\ 2p_z & 1s_z \end{bmatrix}
\begin{bmatrix} 2s & 2s \\ 2p_x & 2p_y \\ 1s_x & 2p_z \\ 1s_y & 1s_z \end{bmatrix}
\begin{bmatrix} 2s & 2s \\ 2p_x & 1s_y \\ 1s_x & 2p_z \\ 2p_y & 1s_z \end{bmatrix},
$$

and these will be symbolized by $T_1, \ldots, T_5$ in the order given. The $A_1$ projector for $C_{3v}$ is

$$e^{A_1} = \frac{1}{6}(I + C_3 + C_3^2 + \sigma_x + \sigma_y + \sigma_z),$$ (6.21)

and using symgenn from the CRUNCH suite, we find that $\theta\mathcal{NPN}T_1$ is a $^1A_1$ symmetry function on its own,

$$e^{A_1}\theta\mathcal{NPN}T_1 = \theta\mathcal{NPN}T_1.$$ (6.22)

Applying $e^{A_1}$ to $\theta\mathcal{NPN}T_2$ yields

$$e^{A_1}\theta\mathcal{NPN}T_2 = \frac{1}{6}\theta\mathcal{NPN}(2T_2 + 2T_3 - T_4 + 3T_5).$$ (6.23)

Using $e^{A_1}$ with $T_3$, $T_4$, or $T_5$ does not give a function linearly independent of those we have found already. Thus, there are two linearly independent $^1A_1$ functions that can be formed from the configuration above. The first of these is not hard to understand when one examines the consequences of the antisymmetry of the columns of the standard tableaux functions. The second, however, is much less obvious and would be very tedious to determine without the computer program.

To obtain the symmetry functions in terms of HLSP functions we can transform the standard tableaux functions using the methods of Chapter 5. The transformation matrix is given in Eq. (5.128):

$$A = \begin{bmatrix} 0 & 0 & 0 & 0 & -2/3 \\ -1/3 & -1/3 & 1/3 & 1/3 & -1/3 \\ -1/3 & 1/3 & -1/3 & 1/3 & -1/3 \\ 0 & 2/3 & 2/3 & 2/3 & -2/3 \\ -1/3 & 1/3 & 1/3 & -1/3 & -1 \end{bmatrix},$$ (6.24)

and multiplying this by the coefficients of the symmetry functions of Eqs. (6.22) and (6.23), we obtain

$$A \begin{bmatrix} 1 & 0 \\ 0 & 1/3 \\ 0 & 1/3 \\ 0 & -1/6 \\ 0 & 1/2 \end{bmatrix} = \begin{bmatrix} 0 & -1/3 \\ -1/3 & -2/9 \\ -1/3 & -2/9 \\ 0 & 0 \\ -1/3 & -2/9 \end{bmatrix},$$ (6.25)

as the coefficients of $^1A_1$ symmetry HLSP functions. The Rumer tableaux, $T_i^{(R)}$, for the HLSP functions are

$$
\begin{bmatrix} 2s & 2s \\ 2p_x & 1s_x \\ 2p_y & 1s_y \\ 2p_z & 1s_z \end{bmatrix}_R
\begin{bmatrix} 2s & 2s \\ 2p_x & 1s_x \\ 1s_y & 2p_z \\ 2p_y & 1s_z \end{bmatrix}_R
\begin{bmatrix} 2s & 2s \\ 1s_x & 2p_y \\ 2p_x & 1s_y \\ 2p_z & 1s_z \end{bmatrix}_R
\begin{bmatrix} 2s & 2s \\ 1s_x & 2p_y \\ 1s_y & 2p_z \\ 2p_x & 1s_z \end{bmatrix}_R
\begin{bmatrix} 2s & 2s \\ 2p_y & 1s_y \\ 1s_x & 2p_z \\ 2p_x & 1s_z \end{bmatrix}_R
$$

These Rumer tableaux are based upon the following diagrams:

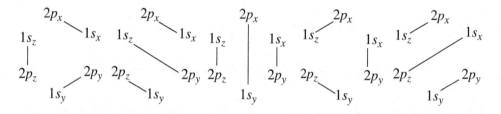

where we have arranged[3] the orbitals below the $2s$ pair in a circle. $T_1^{(R)}$ and $T_4^{(R)}$ are the two "Kekulé"[4] diagrams and the others are the "Dewar" diagrams. $T_1^{(R)}$ is the HLSP function with three electron pair bonds between the $2p_i$ orbital and the closest $2s_i$. One sees that the $T_4^{(R)}$ Kekulé structure is completely missing from the $^1A_1$ functions. We, of course, could have determined the symmetry HLSP functions by examining them directly. Clearly $T_1^{(R)}$ is by itself a symmetry function and a sum of the three Dewar structures is also. It is not so obvious that $T_4^{(R)}$ does not contribute.

One must confess that these $A_1$ symmetry results we have obtained for NH$_3$ are reasonably simple, because we chose the order of the AOs the way we did. One could arrange the orbitals in some other order and obtain valid results, but have symmetry functions that are very nonintuitive. The reader is urged to experiment with `symgenn` to see this.

This is also evident when we consider $A_2$ symmetry, the projector for which is

$$
e^{A_2} = \frac{1}{6}\left(I + C_3 + C_3^2 - \sigma_x - \sigma_y - \sigma_z\right). \tag{6.26}
$$

As $e^{A_2}$ is applied to the $T_i$ in turn, we obtain zero until

$$
e^{A_2}\theta\mathcal{NPN}T_4 = \frac{1}{2}\theta\mathcal{NPN}(T_4 - T_5). \tag{6.27}
$$

---

[3] The order of these Rumer diagrams is determined by the automatic generation routine in the computer program.

[4] Although NH$_3$ does not have the spatial symmetry of a hexagon, we still may use this terminology in describing the structures.

$T_5$ gives the same function. This function may be written in terms of Rumer tableaux, $T_i^{(R)}$ also, and we obtain

$$\frac{1}{2}\theta\mathcal{NPN}(T_4 - T_5) = \frac{1}{9}\mathcal{NP}\big(T_1^{(R)} + T_2^{(R)} + T_3^{(R)} + 2T_4^{(R)} + T_5^{(R)}\big), \quad (6.28)$$

not a result that is particularly intuitive. $T_4^{(R)}$ is present in this sum, and we will see shortly that this Rumer tableau is the only one that has an $A_2$ component. It is not pure, as is seen from Eq. (6.28), but none of the other tableaux have any $A_2$ part at all.

There are two linear combinations of the standard tableaux functions that comprise a pair of $E$ symmetry. The $E$ projectors are

$$e_{11}^E = \frac{1}{3}\big(I - \tfrac{1}{2}C_3 - \tfrac{1}{2}C_3^2 + \sigma_x - \tfrac{1}{2}\sigma_y - \tfrac{1}{2}\sigma_z\big), \quad (6.29)$$

$$e_{21}^E = \frac{1}{3}\big(\sqrt{3}/_2 C_3 - \sqrt{3}/_2 C_3^2 + \sqrt{3}/_2\sigma_y - \sqrt{3}/_2\sigma_z\big). \quad (6.30)$$

The computations show that

$$e_{11}^E \theta\mathcal{NPN}T_2 = \frac{1}{3}\theta\mathcal{NPN}(2T_2 - T_3 - T_4), \quad (6.31)$$

$$e_{21}^E \theta\mathcal{NPN}T_3 = \frac{\sqrt{3}}{6}\theta\mathcal{NPN}(-T_3 + T_4), \quad (6.32)$$

where the energy for either component will be the same. Again, the functions may be expressed in terms of the Rumer structures, and we obtain

$$\frac{1}{3}\theta\mathcal{NPN}(2T_2 - T_3 - T_4) = \frac{2}{9}\mathcal{NP}\big(-2T_2^{(R)} + T_3^{(R)} + T_5^{(R)}\big), \quad (6.33)$$

$$\frac{\sqrt{3}}{6}\theta\mathcal{NPN}(-T_3 + T_4) = \frac{1}{\sqrt{27}}\mathcal{NP}\big(T_3^R - T_5^R\big). \quad (6.34)$$

As we commented on above, $T_4^R$ is missing from all of the functions except for the one of $A_2$ symmetry.

### 6.3.3 Example 3. The π system of benzene

In Chapter 15 we give an extensive treatment of the π system of benzene, but here we outline briefly some of the symmetry considerations. We consider the configuration $p_1 p_2 p_3 p_4 p_5 p_6$, where $p_i$ stands for a $C2p_z$ orbital at the $i^{th}$ C atom, numbered sequentially and counterclockwise around the ring. The set of five standard tableaux is

$$\begin{bmatrix} p_1 & p_2 \\ p_3 & p_4 \\ p_5 & p_6 \end{bmatrix} \begin{bmatrix} p_1 & p_2 \\ p_3 & p_5 \\ p_4 & p_6 \end{bmatrix} \begin{bmatrix} p_1 & p_3 \\ p_2 & p_4 \\ p_5 & p_6 \end{bmatrix} \begin{bmatrix} p_1 & p_3 \\ p_2 & p_5 \\ p_4 & p_6 \end{bmatrix} \begin{bmatrix} p_1 & p_4 \\ p_2 & p_5 \\ p_3 & p_6 \end{bmatrix}$$

and we label them $T_1, \ldots, T_5$ in that order. The $e^{A_{1g}}$ in merely the sum of all of the elements of the $D_{6h}$ symmetry group divided by 24, the value of $g$. We obtain two linear combinations

$$e^{A_{1g}}\theta\mathcal{NPN}T_1 = \theta\mathcal{NPN}T_1, \tag{6.35}$$

$$e^{A_{1g}}\theta\mathcal{NPN}T_2 = \frac{1}{6}\theta\mathcal{NPN}(3T_1 - T_2 - T_3 + 2T_4 - 3T_5). \tag{6.36}$$

Here again, the second of these is not obviously a symmetry function.

The Rumer diagrams for benzene actually mirror the real spatial symmetry, and thus the Kekulé and Dewar structures emerge,

and with these we associate the Rumer tableaux

$$
\begin{bmatrix} p_1 & p_2 \\ p_3 & p_4 \\ p_5 & p_6 \end{bmatrix}_R
\begin{bmatrix} p_1 & p_2 \\ p_3 & p_6 \\ p_4 & p_5 \end{bmatrix}_R
\begin{bmatrix} p_1 & p_4 \\ p_2 & p_3 \\ p_5 & p_6 \end{bmatrix}_R
\begin{bmatrix} p_1 & p_6 \\ p_2 & p_3 \\ p_4 & p_5 \end{bmatrix}_R
\begin{bmatrix} p_1 & p_6 \\ p_2 & p_5 \\ p_3 & p_4 \end{bmatrix}_R .
$$

The transformation from standard tableaux functions to HLSP functions is independent of the spatial symmetry and so we need the $A$-matrix in Eq. (6.24) again. This time the results are

$$
A\begin{bmatrix} 1 & 1/2 \\ 0 & -1/6 \\ 0 & -1/6 \\ 0 & 1/3 \\ 0 & -1/2 \end{bmatrix} = \begin{bmatrix} 0 & 1/3 \\ -1/3 & 2/9 \\ -1/3 & 2/9 \\ 0 & 1/3 \\ -1/3 & 2/9 \end{bmatrix}. \tag{6.37}
$$

In this case the symmetry functions in terms of the Rumer tableau are fairly obvious, as can be seen by inspection of the Rumer diagrams added together in them.

We give more details of symmetry in benzene in Chapter 15.

# 7

# Varieties of VB treatments

The reader will recall that in Chapter 2 we gave examples of $H_2$ calculations in which the orbitals were restricted to one or the other of the atomic centers and in Chapter 3 the examples used orbitals that range over more than one nuclear center. The genealogies of these two general sorts of wave functions can be traced back to the original Heitler–London approach and the Coulson–Fisher[15] approach, respectively. For the purposes of discussion in this chapter we will say the former approach uses *local orbitals* and the latter, *nonlocal orbitals*. One of the principal differences between these approaches revolves around the occurrence of the so-called ionic structures in the local orbital approach. We will describe the two methods in some detail and then return to the question of ionic structures in Chapter 8.

## 7.1 Local orbitals

The use in VB calculations of local orbitals is more straightforward than the alternative. In its simplest form, when atomic AOs are used and considered fixed, the wave function is

$$\Psi = \sum_i C_i \phi_i, \qquad (7.1)$$

where the $\phi_i$ are $n$-electron basis functions as described in Chapter 5. The wave function presents a linear variation problem, and the only real problem is the practical one of choosing a suitable set of $\phi_i$ functions. We will discuss this latter problem more fully in Chapter 9.

A primary characteristic of this approach is that each $\phi_i$ can be interpreted as a representation of the molecule in which each atom has a more-or-less definite state or configuration. In this way the molecule as a whole may be thought of as consisting of a mixture of atomic states including ionic ones, and in ideal circumstances we may calculate fractional weights for these states. The focus here is thus on the way atoms in a number of states interact to form the molecule.

This is, of course, the approach used by all of the early VB workers. In more recent times, after computing machinery allowed *ab initio* treatments, this is the sort of wave function proposed by Balint-Kurti and Karplus[34], which they called a *multistructure approach*. The present author and his students have proposed the multiconfiguration valence bond (MCVB) approach, which differs from the Balint-Kurti–Karplus wave function principally in the way the $\phi_i$ are chosen.

The local approach may be extended, as Hiberty[44] suggests, by allowing the AOs to "breathe". This is accomplished in modern times by writing the orbitals in $\phi_i$ as linear combinations of more primitive AOs, all at one nuclear center, and optimizing these linear combinations along with the coefficients in Eq. (7.1). The breathing thus contributes a *nonlinear* component to the energy optimization. This latter is, of course, only a practical problem; it contributes no conceptual difficulty to the interpretation of the wave function.

We may summarize the important characteristics of VB calculations with local orbitals.

1. The $n$-electron basis consists of functions that have a clearcut interpretation in terms of individual atomic states or configurations.
2. Many atomic states in $\phi_i$ are of the sort termed "ionic".
3. In a highly accurate energy calculation many terms may be required in Eq. (7.1).
4. If Rumer tableaux are used for $\phi_i$, these may in many cases be put in a one-to-one relation with classical bonding diagrams used by chemists.
5. In its simplest form the energy optimization is a linear variation problem.
6. If a molecule dissociates, the asymptotic wave function has a clear set of atomic states.

## 7.2 Nonlocal orbitals

In all of the various VB methods that have been suggested involving nonlocal orbitals it is obvious that the orbitals must be written as linear combinations of AOs at many centers. Thus one is always faced with some sort of nonlinear minimization of the Rayleigh quotient.

Historically, the first of the modern descendents of the Coulson–Fisher method proposed was the GGVB approach. Nevertheless, we will postpone its description, since it is a restricted version of still later proposals.

We describe first the SCVB proposal of Gerratt *et al.* We use here the notation and methods of Chapter 5. These workers originally used the *genealogical representations* of the symmetric groups[7], but so long as the irreducible representation space is completely spanned, any representation will give the same energy and wave function. Balint-Kurti and van Lenthe proposed using an equivalent wave function. The principal differences between these proposals deal with methods of optimization. We will continue to use the SCVB acronym for this method.

Consider a system of $n$ electrons in a spin state $S$. We know that there are for $n$ linearly independent orbitals

$$f = \frac{2S+1}{n+1} \binom{n+1}{n-S/2}$$

(7.2)

linearly independent standard tableaux functions or HLSP functions that can be constructed from these orbitals. In the present notation the SCVB wave function is written as the general linear combination of these:

$$\Psi_{SCVB} = \sum_{i=1}^{f} C_i \phi_i(u_1, \ldots, u_n),$$

(7.3)

where the orbitals in $\phi_i$ are, in general, linear combinations of the whole AO basis.[1] The problem is to optimize the Rayleigh quotient for this wave function with respect to both the $C_i$ and the linear coefficients in the orbitals. Using familiar methods of the calculus of variations, one can set the first variation of the energy with respect to the orbitals and linear coefficients to zero. This leads to a set of Fock-like operators, one for each orbital. Gerratt *et al.* use a second order stabilized Newton–Raphson algorithm for the optimization. This gives a set of occupied and virtual orbitals from each Fock operator as well as optimum $C_i$s.

The SCVB energy is, of course, just the result from this optimization. Should a more elaborate wave function be needed, the virtual orbitals are available for a more-or-less conventional, but nonorthogonal, CI that may be used to improve the SCVB result. Thus an accurate result here may also involve a wave function with many terms.

The GGVB[41] wave function can have several different forms, each one of which, however, is a restricted version of a SCVB wave function. As originally proposed, a GGVB calculation uses just one of the genealogical irreducible representation functions and optimizes the orbitals in it, under a constraint of some orthogonality. In general, the orbitals are ordered into two sets, orthogonality is enforced within the sets but not between them. Thus, there are $f$ different GGVB wave functions, depending upon which of the genealogical $\phi_i$ functions is used. Goddard designated these as the G1, G2, ..., Gf methods, the general one being Gi. Each of these, in general, yields a different energy, and one could choose the wave function for the lowest as the best GGVB answer. In actual practice only the G1 or Gf methods have been much used. In simple cases the Gf wave function is a standard tableaux function while the G1 is a HLSP function. For Gf wave functions

---

[1] The requirements of symmetry may modify this.

we may show that the above orthogonality requirement is not a real constraint on the energy. From Chapter 5 we have, for $n$ electrons in a singlet state, the unnormalized function

$$\Psi_{Gf} = \theta \mathcal{N} \mathcal{P} \mathcal{N} u_1(1) \cdots u_m(m) v_1(m+1) \cdots v_m(n), \qquad (7.4)$$

where $m = n/2$, and the $u$s and $v$s are the two sets of orbitals. In principle, we could optimize the energy and orbitals corresponding to Eq. (7.4), and afterwards the presence of the $\mathcal{N}$ will allow the formation of linear combinations among the $u$s and, in general, different ones among the $v$s that will render the two sets internally orthonormal. This does not change the value of $\Psi_{Gf}$, of course, except possibly for its overall normalization.

On the other hand, no such invariance of G1 or HLSP functions occurs, so the orthogonality constraint has a real impact on the calculated energy.

We saw in Chapter 3 how the delocalization of the orbitals takes the place of the ionic terms in localized VB treatments, and this phenomenon is generally true for $n$ electron systems.

We now summarize the main characteristics of VB calculations with nonlocal orbitals.

1. The wave function is reasonably compact, normally having no more than $f$ terms.
2. There are no structures in the sum that must be interpreted as "ionic" in character. For many people this is a real advantage to these VB functions.
3. The SCVB function produces a considerable portion of the correlation energy.
4. If Rumer tableaux are used for $\phi_i$, these may in many cases be put in a one-to-one relation with classical bonding diagrams used by chemists.
5. If a molecule dissociates, the asymptotic wave function has a clear set of atomic states.

Illustrations of both of these classes of VB functions will be given for a number of systems in Part II of this book.

# 8

# The physics of ionic structures

The existence of many ionic structures in MCVB wave functions has often been criticized by some workers as being unphysical. This has been the case particularly when a covalent bond between like atoms is being represented. Nevertheless, we have seen in Chapter 2 that ionic structures contribute to electron delocalization in the $H_2$ molecule and would be expected to do likewise in all cases. Later in this chapter we will see that they can also be interpreted as contributions from ionic states of the constituent atoms. When the bond is between unlike atoms, it is to be expected that ionic structures in the wave function will also contribute to various electric moments, the dipole moment being the simplest. The amounts of these ionic structures in the wave functions will be determined by a sort of "balancing act" in the variation principle between the "diagonal" effects of the ionic state energies and the "off-diagonal" effect of the delocalization.

In this chapter we will also focus on the dipole moment of molecules. With these, some of the most interesting phenomena are the molecules for which the electric moment is in the "wrong" direction insofar as the atomic electronegativities are concerned. CO is probably the most famous of these cases, but other molecules have even more striking disagreements. One of the larger is the simple diatomic BF. We will take up the question of the dipole moments of molecules like BF in Chapter 12. In this chapter we will examine in a more general way how various sorts of structures influence electric moments for two simple cases. For some of the discussion in this chapter we restrict ourselves to descriptions of minimal basis set results, since these satisfactorily describe the physics of the effects. In other cases a more extensive treatment is necessary.

## 8.1 A silly two-electron example

In Chapter 2 we described several treatments of the $H_2$ molecule, and, of course, there the question of dipole moments was irrelevant, although we could have calculated a quadrupole moment. We want here to consider the properties of a covalent VB structure for $H_2$ using a silly basis set consisting of $1s$ orbitals for the two atoms that have different scale factors. Such a wave function is certain not to have the correct $^1\Sigma_g^+$ symmetry for $H_2$ and will not have a credible energy, but an important point emerges. Let the singlet standard tableaux function, $\psi$, be

$$\psi = N [1s_a' \ 1s_b], \tag{8.1}$$

where, as indicated by the prime, we use differently scaled AOs at the two centers. The question we ask is: What is the electric dipole moment implied by this wave function? Assuming the molecule is situated along the $z$-axis, the $x$- and $y$-components of the moment are zero. In atomic units the $z$-component of the moment is

$$\mu_z = z_a + z_b - \frac{z_a + z_b + 2S\langle 1s_a'|z|1s_b\rangle}{1 + S^2},$$
$$= \frac{S[S(z_a + z_b) - 2\langle 1s_a'|z|1s_b\rangle]}{1 + S^2}, \tag{8.2}$$

where $S = \langle 1s_a'|1s_b\rangle$. It is clear from Eq. (8.2) that, whatever its value at small distances, $\mu_z$ goes to zero as the interatomic distance goes to infinity, since $S$ also goes to zero. $\mu_z$ is not zero, however, at 0.7 Å, a distance near that at equilibrium in $H_2$. Taking the scale of $1s$ to be 1.0 and that of $1s'$ to be 1.2, a value close to that which is optimum for the molecule,[1] we obtain $\mu_z = -0.118$ D. STO6Gs were used and $z_b < z_a$, i.e., the less diffuse orbital is in the positive $z$-direction from the other. This calculated moment is not very large, but it arises from a *purely covalent* function. If we do the same calculation for the triplet function,

$$\begin{bmatrix} 1s_a' \\ 1s_b \end{bmatrix},$$

we obtain $\mu_z = 0.389$ D, in the direction opposite to that for the singlet function. In the singlet case the electron distribution is more toward the less extended AO and in the triplet case more toward the more extended AO.

It is useful to state this result in different language. In general, we expect more electronegative atoms to have tighter less diffuse orbitals in comparable shells than atoms of lower electronegativity. In our case this means we have a surrogate atom for higher electronegativity in the positive $z$-direction from the other. Therefore, bonding interactions have the electrons moving toward the more electronegative atom and antibonding interactions have them moving toward the less electronegative atom. The usual sign convention confusion occurs, of course; the dipole moment points in a direction opposite to the electron movement.

---

[1] We note that if the orbitals were scaled equally $2\langle 1s_a|z|1s_b\rangle = S(z_a + z_b)$ and $\mu_z$ is correctly zero at all distances.

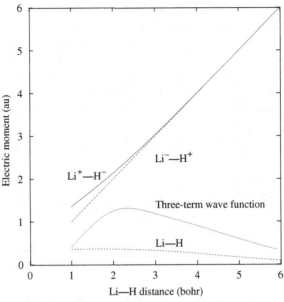

Figure 8.1. The electric dipole function versus the internuclear distance for LiH using the two ionic and one covalent functions, individually. The sign of the moment for $Li^- - H^+$ has been changed to facilitate plotting and comparison of the magnitudes.

It is clear, when we calculate the electric moment with a more realistic wave function, that even the so-called covalent functions can make nonzero contributions when symmetry allows.

## 8.2 Ionic structures and the electric moment of LiH

There are two rather different questions that arise when considering ionic structures in VB wave functions. The first of these we discuss is the contribution to electric dipole moments. LiH is considered as an example. In the next section we take up ionic structures and curve crossings, using LiF to illustrate the points.

LiH is the simplest uncharged molecule that has a permanent electric dipole moment.[2] We examine here some of the properties of the simplest VB functions for this molecule. The molecule is oriented along the $z$-axis with the Li atom in the positive direction.

Figure 8.1 shows the expectation values of the electric moment for the $Li^+ - H^-$ and covalent structures. The graph gives the *negative* of the moment for $Li^- - H^+$ for easy comparison. In addition, the moment for the three-term wave function involving all three of the other functions is given. Although such a simple wave

---

[2] The small moments in isotopic hydrogen, HD and HT, for example, do not interest us here.

function does not reproduce the total moment accurately, qualitatively the signs are correct. The behavior of the three expectation values deserves comment.

1. $Li^- - H^+$ From the point of view of electrostatics this structure is the simplest. We have a spherical $Li^-$ ion and a bare proton. The dipole moment should just be $-R$ in atomic units, and the line giving the dependence of the moment on $R$ should be straight with a slope of $-45°$ and through the origin. (NB The sign of this curve has been changed in Fig. 8.1 to facilitate comparison of magnitudes.)

2. $Li^+ - H^-$ At longer distances the molecule is essentially an undistorted $Li^+$ ion and an $H^-$ ion. As such, the moment equals the internuclear distance. As the ions approach one another, the Pauli principle interaction between the Li$1s$ and H$1s$ AOs causes the moment to be larger than the value due only to the distance. It should be noted that the effect of the Pauli principle is in the same direction as is the triplet example we gave earlier for the $H_2$ molecule with unequally scaled AOs, i.e., the charge density is pushed toward the more diffuse orbital. The exchange interaction between two doubly occupied orbital distributions is essentially like the triplet interaction between two singly occupied orbitals of the same sort.

3. Covalent The dipole moment of the covalent structure is never larger than 1 (in atomic units) and is always positive. A simple analysis is not so easy here, but the same Li$1s$ and H$1s$ interaction as appeared above occurs in this case also.

Figure 8.2 shows the coefficients of the three structures in the total three-term wave function. As expected, the covalent term predominates at all distances, but

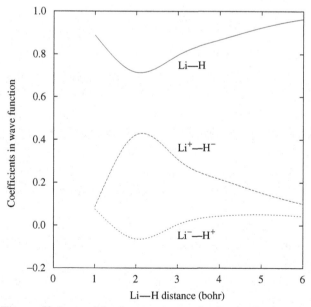

Figure 8.2. The coefficients of the three structures in the simple three-term wave function.

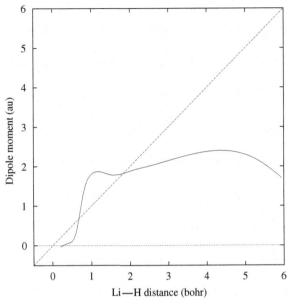

Figure 8.3. The dipole function for the full valence wave function of LiH. It is a little difficult to see on the present scale, but the moment is –0.033 au at an internuclear distance of 0.2 bohr.

the $Li^+–H^-$ term is the next largest at all distances. The structure with the "wrong" sign is always the smallest in the range of the graph. Figure 8.2 admittedly does not allow a quantitative analysis of the way the three terms produce the total moment, but it does provide a suggestive picture of how the various terms contribute.

When a full valence calculation is done with the present basis, there are 48 standard tableaux functions that produce 45 $^1\Sigma^+$ functions. Figure 8.3 shows the dipole moment function for this wave function. The figure also shows the "45°" line that would be the moment if the charges at the ends were unit magnitude. The curve has some interesting structure. As noted in the caption, even this molecule shows the dipole going the "wrong" way to a slight extent at very close internuclear distances. As the distance increases the moment rises above the 45° line, indicating an effective charge on the ends greater in magnitude than one au. By the time the equilibrium distance is reached (3.019 bohr), the effective charges have fallen so that the moment is 5.5 D with the positive end at the Li, as the electronegativities predict.

## 8.3 Covalent and ionic curve crossings in LiF

Lithium and fluorine form a diatomic molecule that has a large dipole moment in the gas phase; it has been measured to be 6.3248 D in the ground vibrational state. The equilibrium internuclear distance is 1.564 Å, and, therefore, the apparent

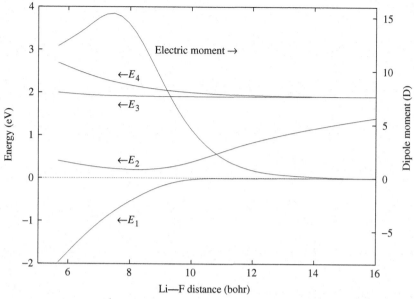

Figure 8.4. The first four $^1\Sigma$ states of LiF near the ionic–covalent curve crossing region. The electric dipole moment is also given showing its change in relation to the crossing of the first two states. The energy curves refer to the left vertical axis and the dipole moment to the right.

charge on the atoms is $\pm 0.84$ electrons. When there is a long distance between them, the overall ground state is the pair of atoms, each in its neutral ground state. The first excited overall state has the Li atom excited $2s \rightarrow 2p$ at 1.847 eV. The ionization potential of Li is 5.390 eV and the electron affinity of F is 3.399 eV, so the next state up is the transfer of an electron from Li to F at an energy of 1.991 eV. This is a lower energy than any $2s \rightarrow 3l$ states of Li or any $2p \rightarrow 3s$ states of F.

If there were no other interaction the energy of the atoms in the $Li^+ - F^-$ state would fall as they approach each other from $\infty$. The energy of this state would cross the ground state at about 12 bohr. In actuality the states interact, and there is an avoided crossing. The energies of the first four states of LiF are shown in Fig. 8.4 as a function of internuclear distance. In addition, the electric dipole moment is shown.

The point we wish to bring out here is that the dipole moment curve around 6–7 bohr is very nearly the value expected for $\pm$ one electronic charge separated by that distance. As one moves outward, the avoided crossing region is traversed, the state of the molecule switches over from $Li^+ - F^-$ to Li–F and the dipole falls rapidly.

The avoided crossing we have discussed occurs between the two curves in Fig. 8.4 labeled $E_1$ and $E_2$. Another avoided crossing, farther out than our graph shows, occurs between the $Li^+-F^-$ state curve and $E_3$ and $E_4$. These latter two states involve $Li2p_\sigma$–$F2p_\sigma$ and $Li2p_\pi$–$F2p_\pi$, both coupled to $^1\Sigma^+$. The crossing in these cases occurs at such large distances there is very little interaction.

The calculations in this illustration were not done with a minimal basis set, since, if such were used, they would not show the correct behavior, even qualitatively. This happens because we must represent both F and $F^-$ in the same wave function. Clearly one set of AOs cannot represent both states of F. Li does not present such a difficult problem, since, to a first approximation, it has either one orbital or none. The calculations of Fig. 8.4 were done with wave functions of 1886 standard tableaux functions. These support 1020 $^1\Sigma^+$ symmetry functions. We will discuss the arrangement of bases more fully in Chapter 9.

# Part II

Examples and interpretations

# 9

# Selection of structures and arrangement of bases

Since, for any but the smallest of systems, a full VB calculation is out of the question, it is essential to devise a useful and systematic procedure for the arrangement of the bases and for the selection of a manageable subset of structures based upon these orbitals. These two problems are interrelated and cannot be discussed in complete isolation from one another, but we will consider the basis question first. In our two-electron calculations we have already addressed some of the issues, but here we look at the problems more systematically.

## 9.1 The AO bases

The calculations described in this section of the book have, for the most part, been carried out using three of the basis sets developed by the Pople school.

**STO3G** A minimal basis. This contains exactly the number of orbitals that might be occupied in each atomic shell.
**6-31G** A valence double-$\zeta$ basis. This basis has been constructed for atoms up through Ar.
**6-31G\*** A valence double-$\zeta$ basis with polarization functions added. Polarization functions are functions of one larger $l$-value than normally occurs in an atomic shell in the ground state.

Any departures from these will be spelled out at the place they are used.

Our general procedure is to represent the atoms in a molecule using the Hartree–Fock orbitals of the individual atoms occurring in the molecule. (We will also consider the interaction of molecular fragments where the Hartree–Fock orbitals of the fragments are used.) These are obtained with the above bases in the conventional way using Roothaan's RHF or ROHF procedure[45], extended where necessary.

ROHF calculations are not well defined, and the reader is cautioned that this term has meanings that differ among workers. Some computational packages, GAMESS[46] is an example, in doing a single-atom calculation, do not treat the atom in a spherical environment. For example, for N in its $^4S$ ground state, the three $p$ orbitals are divided into $\sigma$ and $\pi$ sets and are not all equal. This is not a matter of any importance in that milieu. An atom in a molecule will not be in a spherically symmetric environment for the Hartree–Fock function to be determined.

In all of the calculations described here, however, we use the original Roothaan specifications that produce sets of $l$-functions that transform into one another under all rotations. This can have an important consequence in our VB calculations, if we treat a problem in which the energies of the system are important as we move to asymptotic geometries. An example will clarify this point. $C_2$ is in a $^1\Sigma_g^+$ ground state, but there are two couplings of two C atoms, each in a $^3P^e$ ground state, that have this symmetry. In our calculations these two will have the correct asymptotic degeneracy only if we use "spherical" atoms.

Conventional basis set Hartree–Fock procedures also produce a number of *virtual orbitals* in addition to those that are occupied. Although there are experimental situations where the virtual orbitals can be interpreted physically[47], for our purposes here they provide the necessary fine tuning of the atomic basis as atoms form molecules. The number of these virtual orbitals depends upon the number of orbitals in the whole basis and the number of electrons in the neutral atom. For the B through F atoms from the second row, the minimal STO3G basis does not produce any virtual orbitals. For these same atoms the 6-31G and 6-31G* bases produce four and nine virtual orbitals, respectively. There is a point we wish to make about the orbitals in these double-$\zeta$ basis sets. A valence orbital and the corresponding virtual orbital of the same $l$-value have approximately the same extension in space. This means that the virtual orbital can efficiently correct the size of the more important occupied orbital in linear combinations. As we saw in the two-electron calculations, this can have an important effect on the AOs as a molecule forms. We may illustrate this situation using N as an example.

The 6-31G basis for N has three $s$-type Gaussian groups. In the representation of the normal atom the $1s$ and $2s$ occupied orbitals are two linear combinations of the three-function basis and the $s$-type virtual orbital is the third. For convenience, we will call the last orbital $3s$, but it should not be thought to be a good representation of a real orbital of that sort in an excited atom. A typical Hartree–Fock calculation yields

$$1s = 0.996\,224\,75 g_6 + 0.019\,984\,19 g_3 - 0.004\,639\,97 g_1, \tag{9.1}$$

$$2s = -0.226\,268\,21 g_6 + 0.515\,913\,17 g_3 + 0.568\,841\,00 g_1, \tag{9.2}$$

$$3s = 0.091\,019\,32 g_6 - 1.5148\,075\,3 g_3 + 1.478\,256\,46 g_1, \tag{9.3}$$

where $g_6$, $g_3$, and $g_1$ are, respectively, the 6, 3, and 1 Gaussian groups from the basis. The $1s$ orbital is predominantly the $g_6$ function, but the other two have roughly equal (in magnitude) parts of $g_3$ and $g_1$. Therefore, a linear combination, $a2s + b3s$ with a fairly small $b$ can have an extension in space differing from that of $2s$ itself. If $b/a < 0$ it is more compact; if $b/a > 0$ it is less compact.

## 9.2 Structure selection

Our discussion of the structure selection must be somewhat more involved. In part, this is a discussion of a crucial member of the CRUNCH package, the program entitled symgenn. It possesses a number of configuration selection devices, for the details of which the reader is referred to the CRUNCH manual. The present discussion will focus on the desired outcome of the selections rather than on how to accomplish them. Again, it is convenient to describe these by giving an example, that of the $N_2$ molecule, which will be discussed quantitatively in Chapter 11.

### 9.2.1 N₂ and an STO3G basis

$N_2$ has 14 electrons and there are 10 orbitals in an STO3G basis. The Weyl dimension formula, Eq. (5.115), gives 4950 configurations[1] for a singlet state. Physical arguments suggest that configurations with electrons excited out of the $1s$ cores should be quite unimportant. If we force a $1s_a^2 1s_b^2$ occupation at all times, Eq. (5.115) now gives us 1176 configurations, a considerable reduction. These are not just $\Sigma_g^+$ states, of course. Symgenn will allow us to select linear combinations having this spatial symmetry only. This reduces the size of the linear variation matrix to $102 \times 102$, a further significant reduction. Another number that symgenn tells us is that, among the 1176 configurations, only 328 appear as any part of a linear combination giving a $\Sigma_g^+$ state. This number would be difficult to determine by hand.[2]

At this stage symgenn has done its job and the matrix generator uses the symgenn results to compute the Hamiltonian matrix. Thus, we would call this a *full valence* calculation of the energy of $N_2$ with an STO3G basis.

### 9.2.2 N₂ and a 6-31G basis

We still have 14 electrons, but the larger basis provides 18 orbitals in the basis. The full calculation now has 4 269 359 850 configurations, a number only slightly

---

[1] This is the number of linearly independent standard tableaux or Rumer functions that the entire basis supports.

[2] To be precise, we should point out that we have symgenn treat $N_2$ as a $D_{4h}$ system, rather than the completely correct $D_{\infty h}$. In projecting symmetry blocks out of Hamiltonian matrices, it is never wrong to use a *subgroup* of the full symmetry, merely inefficient. It would be a serious error, of course, to use too high a symmetry. It happens for the STO3G basis that there is no difference between $A_{1g}$ $D_{4h}$ and $\Sigma_g^+$ $D_{\infty h}$ projections.

smaller than $2^{32}$. Forcing a $1s_a^2 1s_b^2$ occupation reduces this to 4 504 864, which is a considerable reduction, but still much too large a number in practice. The reduction of these to $\Sigma_g^+$ states is not known, but the number is still likely to be considerable. Instead, we use physical arguments again to reduce the number of configurations further. Many of the 4 504 864 configurations have mostly virtual orbitals occupied and we expect these to be unimportant. The number of occupied orbitals from the 6-31G basis is the same number as the total number of orbitals from the STO3G basis. Therefore, there are again 102 $\Sigma_g^+$ functions from the occupied orbitals. These include charge separations as high as $\pm 3$. We add to this full valence set those configurations that have one occupied orbital replaced by one virtual orbital in the valence configurations with charge separation no higher than $\pm 1$. Symgenn could work out the number of configurations resulting, but we have not done this. If this selection scheme is combined with $\Sigma_g^+$ symmetry projection, we obtain a $1086 \times 1086$ Hamiltonian matrix, an easily manageable size.

### 9.2.3 $N_2$ and a 6-31G* basis

When we add $d$ orbitals to the basis on each atom we have the possibility that polarization can occur. Of course, as far as an atom in the second or third rows is concerned, the $d$ orbitals merely increase the number of virtual orbitals and increase the number of possibilities for substitutions from the normally filled set. We do not give any of the numbers here, but will detail them when we discuss particular examples.

### 9.3 Planar aromatic and $\pi$ systems

In later chapters we give a number of calculations of planar unsaturated systems. Because of the plane of symmetry, the SCF orbitals can be sorted into two groups, those that are even with respect to the symmetry plane, and those that are odd. The former are commonly called $\sigma$ orbitals and the latter $\pi$ orbitals. Although it is an approximation, there has been great interest in treating the $\pi$ parts of these systems with VB methods and ignoring the $\sigma$ parts. The easiest way of doing this, while still using *ab initio* methods, is to arrange all configurations to have all of the occupied $\sigma$ orbitals doubly occupied in the same way. In addition, $\sigma$ virtual orbitals are simply ignored. The $\pi$ AOs may then be used in their raw state or in any linear combinations desired. In this sort of arrangement, the $\pi$ electrons are subjected to what is called the *static-exchange potential (SEP)*[39] of the nuclei and $\sigma$ core. The most important molecules of this sort are the aromatic hydrocarbons, but many examples containing oxygen and nitrogen also exist.

# 10

# Four simple three-electron systems

In this chapter we describe four rather different three-electron systems: the $\pi$ system of the allyl radical, the $He_2^+$ ionic molecule, the valence orbitals of the BeH molecule, and the Li atom. In line with the intent of Chapter 4, these treatments are included to introduce the reader to systems that are more complicated than those of Chapters 2 and 3, but simple enough to give detailed illustrations of the methods of Chapter 5. In each case we will examine MCVB results as an example of localized orbital treatments and SCVB results as an example of delocalized treatments. Of course, for Li this distinction is obscured because there is only a single nucleus, but there are, nevertheless, noteworthy points to be made for that system. The reader should refer back to Chapter 4 for a specific discussion of the three-electron spin problem, but we will nevertheless use the general notation developed in Chapter 5 to describe the results because it is more efficient.

## 10.1 The allyl radical

All of the calculations on allyl radicals are based upon a conventional ROHF treatment with a full geometry optimization using a 6-31G* basis set. The $\sigma$ "core" was used to construct an SEP as described in Chapter 9. The molecule possesses $C_{2v}$ symmetry. The $C_2$ symmetry axis is along the $z$-axis and the nuclei all reside in the $x$–$z$ plane. Thus the "$\pi$" AOs consist of the $p_y$s, $d_{xy}$s, and $d_{yz}$s, of which there are 12 in all for this basis. At each C there is a $2p_y$, a $3p_y$, a $3d_{xy}$, and a $3d_{yz}$. The $2p_y$ is the SCF orbital for the atomic ground state, and the $3p_y$ is the virtual orbital of the same symmetry. Table 10.1 shows for reference the pertinent portions of the $C_{2v}$ character table. We number the $\pi$ orbitals from one end of the molecule and use $2p_1$, $2p_2$, and $2p_3$, remembering that they are all of the $2p_y$ sort. The effect of the $\sigma_{xz}$ and $\sigma_{yz}$ operations of the group is seen to be

$$\sigma_{yz}2p_i = -2p_i, \tag{10.1}$$

$$\sigma_{xz}2p_1 = 2p_3, \tag{10.2}$$

Table 10.1. $C_{2v}$ characters.

| $C_{2v}$ | $I$ | $C_2$ | $\sigma_{xz}$ | $\sigma_{yz}$ |
|----------|-----|-------|---------------|---------------|
| $A_1$    | 1   | 1     | 1             | 1             |
| $A_2$    | 1   | 1     | $-1$          | $-1$          |
| $B_1$    | 1   | $-1$  | 1             | $-1$          |
| $B_2$    | 1   | $-1$  | $-1$          | 1             |

Table 10.2. *Results of 128-function MCVB calculation.*

| | |
|---|---|
| $-116.433\,248\,63$ au | SCF energy |
| $-116.477\,396\,60$ au | MCVB energy |
| $1.201$ eV | Correlation energy |
| $0.9003$ | EGSO pop. of $\begin{bmatrix} 2p_1 & 2p_2 \\ 2p_3 & \end{bmatrix}$ |

$$\sigma_{xz}2p_2 = 2p_2, \tag{10.3}$$

$$\sigma_{xz}2p_3 = 2p_1. \tag{10.4}$$

The effect of the $C_2$ operation is easily determined since $C_2 = \sigma_{xz}\sigma_{yz}$. There is, of course, a completely parallel set of relations for the $3p_y$ set of orbitals. Writing out the corresponding relations for the $3d$ orbitals is left to the interested reader.

### 10.1.1 MCVB treatment

An MCVB calculation with a full set of configurations involving the six $2p_y$ and $3p_y$ orbitals with further configurations involving all possible single excitations out of this set into the $d$-set gives 256 standard tableau functions, which can form 128 $^2A_2$ symmetry functions and a Hamiltonian matrix of the same dimension. Table 10.2 gives several results from the calculation, and we see that there is about 1.2 eV of correlation energy. Because of the static exchange core, all of this is in the $\pi$ system, of course. In addition we see that the EGSO population suggests that the wave function is 90% of the basic VB function with unmodified AOs. This is true, of course, for either standard tableaux functions or HLSP functions.

It is instructive to examine the symmetry of the standard tableaux function of highest EGSO population given in Table 10.2. The effects of the two symmetry planes of $C_{2v}$ on the $2p_i$ orbitals are given above, and, consequently,

$$\sigma_{yz} \begin{bmatrix} 2p_1 & 2p_2 \\ 2p_3 \end{bmatrix} = -\begin{bmatrix} 2p_1 & 2p_2 \\ 2p_3 \end{bmatrix}, \tag{10.5}$$

$$\sigma_{xz} \begin{bmatrix} 2p_1 & 2p_2 \\ 2p_3 \end{bmatrix} = \begin{bmatrix} 2p_3 & 2p_2 \\ 2p_1 \end{bmatrix}, \tag{10.6}$$

$$= -\begin{bmatrix} 2p_1 & 2p_2 \\ 2p_3 \end{bmatrix}. \tag{10.7}$$

It is important to recognize why Eq. (10.7) is true. From Chapter 5 we have

$$\begin{bmatrix} 2p_1 & 2p_2 \\ 2p_3 \end{bmatrix} = \theta \mathcal{N} \mathcal{P} \mathcal{N} 2p_1(1)2p_3(2)2p_2(3), \tag{10.8}$$

except for normalization. Since $\mathcal{N}$ is a column antisymmetrizer, if we interchange $2p_1(1)2p_3(2)$, the sign of the whole function changes, and this standard tableaux function has $^2A_2$ symmetry. The spatial projector for $A_2$ symmetry may be constructed from Table 10.1,

$$e^{A_2} = \frac{1}{4}[I + C_2 - \sigma_{xz} - \sigma_{yz}], \tag{10.9}$$

and we see that

$$e^{A_2} \begin{bmatrix} 2p_1 & 2p_2 \\ 2p_3 \end{bmatrix} = \begin{bmatrix} 2p_1 & 2p_2 \\ 2p_3 \end{bmatrix}. \tag{10.10}$$

The second standard tableaux function

$$\begin{bmatrix} 2p_1 & 2p_3 \\ 2p_2 \end{bmatrix}$$

is not a pure symmetry type; in fact, it is a linear combination of $^2A_2$ and $^2B_2$. Since there cannot be three linearly independent functions from these tableaux, the two $^2A_2$ functions must be the same, and we do not need the second standard tableaux function for this calculation. The $e^{A_2}$ operator may be applied to this tableau to obtain the result in a less formal fashion,

$$e^{A_2} \begin{bmatrix} 2p_1 & 2p_3 \\ 2p_2 \end{bmatrix} = \frac{1}{2} \left( \begin{bmatrix} 2p_1 & 2p_3 \\ 2p_2 \end{bmatrix} - \begin{bmatrix} 2p_3 & 2p_1 \\ 2p_2 \end{bmatrix} \right), \tag{10.11}$$

where we have a nonstandard tableau in the result. Again, the methods of Chapter 5 come to our aid, and we have

$$\begin{bmatrix} 2p_3 & 2p_1 \\ 2p_2 \end{bmatrix} = \left( \begin{bmatrix} 2p_1 & 2p_3 \\ 2p_2 \end{bmatrix} - \begin{bmatrix} 2p_1 & 2p_2 \\ 2p_3 \end{bmatrix} \right), \tag{10.12}$$

Table 10.3. *Results of smaller VB calculations.*

| | |
|---|---|
| $-116.433\,248\,63$ au | SCF energy |
| 32-function MCVB – $d$-functions removed | |
| $-116.470\,007\,69$ au | MCVB energy |
| 1.000 eV | Apparent correlation energy |
| 0.9086 | EGSO pop. of $\begin{bmatrix} 2p1 & 2p2 \\ 2p3 \end{bmatrix}$ |
| 4-Function MCVB – $2p_1, 2p_2, 2p_3$ only | |
| $-116.461\,872\,28$ au | MCVB energy |
| 0.779 eV | Apparent correlation energy |
| 0.9212 | EGSO pop. of $\begin{bmatrix} 2p1 & 2p2 \\ 2p3 \end{bmatrix}$ |
| 2-function VB – $2p_1, 2p_2, 2p_3$ covalent only | |
| $-116.413\,426\,76$ | Energy |

and substituting this result into Eq. (10.11), we obtain

$$e^{A_2} \begin{bmatrix} 2p_1 & 2p_3 \\ 2p_2 \end{bmatrix} = \frac{1}{2} \begin{bmatrix} 2p_1 & 2p_2 \\ 2p_3 \end{bmatrix}. \tag{10.13}$$

Our ability to represent the wave function for allyl as one standard tableaux function should not be considered too important. If we had ordered our $2p$ orbitals differently with respect to particle labels, there are cases where the $^2A_2$ function would require using both standard tableaux functions.

This happens when we consider the most important configuration using HLSP functions. The two Rumer diagrams are shown with dots to indicate the extra electron.

Transforming our wave function to the HLSP function basis,[1] we obtain

$$^2A_2 = 0.41115 \left( \begin{bmatrix} 2p_2 & 2p_1 \\ 2p_3 \end{bmatrix}_R - \begin{bmatrix} 2p_2 & 2p_3 \\ 2p_1 \end{bmatrix}_R \right) + \cdots. \tag{10.14}$$

where we have used Rumer tableaux (see Chapter 5). We emphasize that the EGSO populations are the same regardless of the basis.

In Table 10.3 we give data for smaller calculations of the allyl $\pi$ system. As expected, the MCVB energies increase as fewer basis functions are included, the

---

[1] Details of this sort of calculation are given in the next section.

apparent correlation energy decreasing by about 0.5 eV. In fact, unlike the case with $H_2$, the covalent only VB energy is *above* the SCF energy. This is a frequent occurrence in systems where resonance occurs between equivalent structures. It arises because of the delocalization tendencies of the electrons. We will take this question up in greater detail in Chapter 15 when we discuss benzene.

In actuality, the two smaller correlation energies shown in Table 10.3 are not very significant, since the AO basis is really different from that giving the SCF energy. What is significant is the relative constancy of the EGSO weight for the most important configuration.

Since there are only four terms, we give the whole wave function for the smallest calculation. In terms of standard tableaux functions one obtains

$$
{}^2A_2 = 0.730\,79 \begin{bmatrix} 2p_1 & 2p_2 \\ 2p_3 & \end{bmatrix}
$$

$$
+ 0.140\,64 \left( \begin{bmatrix} 2p_1 & 2p_1 \\ 2p_3 & \end{bmatrix} - \begin{bmatrix} 2p_3 & 2p_3 \\ 2p_1 & \end{bmatrix} \right)
$$

$$
+ 0.139\,95 \left( \begin{bmatrix} 2p_2 & 2p_2 \\ 2p_3 & \end{bmatrix} - \begin{bmatrix} 2p_2 & 2p_2 \\ 2p_1 & \end{bmatrix} \right)
$$

$$
+ 0.061\,32 \left( \begin{bmatrix} 2p_1 & 2p_1 \\ 2p_2 & \end{bmatrix} - \begin{bmatrix} 2p_3 & 2p_3 \\ 2p_2 & \end{bmatrix} \right). \tag{10.15}
$$

The HLSP function form of this wave function is easily obtained with the method of Section 5.5.5,

$$
{}^2A_2 = 0.411\,88 \left( \begin{bmatrix} 2p_2 & 2p_1 \\ 2p_3 & \end{bmatrix}_R - \begin{bmatrix} 2p_2 & 2p_3 \\ 2p_1 & \end{bmatrix}_R \right)
$$

$$
+ \text{three other terms the same as in Eq. (10.15)}. \tag{10.16}
$$

The reader will recall that a given configuration has different standard tableaux functions and HLSP functions if and only if it supports more than one standard tableaux function (or HLSP function).

It will be instructive to detail the calculations leading from Eq. (10.15) to Eq. (10.16). This provides an illustration of the methods of Section 5.5.5.

### 10.1.2 Example of transformation to HLSP functions

The permutations we use are based upon the particle label tableau

$$
\begin{bmatrix} 1 & 3 \\ 2 & \end{bmatrix},
$$

and, therefore,

$$\tfrac{1}{6}\mathcal{NPN} = \tfrac{1}{3}[I - (12) + \tfrac{1}{2}(13) + \tfrac{1}{2}(23)$$
$$- \tfrac{1}{2}(123) - \tfrac{1}{2}(132)], \tag{10.17}$$
$$\tfrac{1}{3}\mathcal{NP} = \tfrac{1}{3}[I - (12) + (13) - (132)]. \tag{10.18}$$

The standard tableaux for the present basis are

$$\begin{bmatrix} 2p_1 & 2p_2 \\ 2p_3 \end{bmatrix} \quad \text{and} \quad \begin{bmatrix} 2p_1 & 2p_3 \\ 2p_2 \end{bmatrix};$$

it should be clear that the permutation yielding the second from the first is $(23)$. Thus, the permutations of the sort defined in Eq. (5.64) are $\{\pi_i\} = \{I, (23)\}$, and we obtain

$$M = \begin{bmatrix} 1 & \tfrac{1}{2} \\ \tfrac{1}{2} & 1 \end{bmatrix}, \tag{10.19}$$

where we have used an $\mathcal{NPN}$ version of Eq. (5.73), and the numbers are obtained from the appropriate coefficient in Eq. (10.17).

The Rumer tableaux may be written

$$\begin{bmatrix} 2p_1 & 2p_2 \\ 2p_3 \end{bmatrix}_R \quad \text{and} \quad \begin{bmatrix} 2p_3 & 2p_2 \\ 2p_1 \end{bmatrix}_R$$

and the $\{\rho_i\}$ set is $\{I, (12)\}$. Thus the matrix $B$ from Eq. (5.126) is

$$B = \begin{bmatrix} 1 & -1 \\ 0 & -1 \end{bmatrix}, \tag{10.20}$$

and $A$ from Eq. (5.128) is

$$A = B^{-1}M = \begin{bmatrix} \tfrac{1}{2} & -\tfrac{1}{2} \\ -\tfrac{1}{2} & -1 \end{bmatrix}. \tag{10.21}$$

We also give the inverse transformation

$$A^{-1} = \begin{bmatrix} \tfrac{4}{3} & -\tfrac{2}{3} \\ -\tfrac{2}{3} & -\tfrac{2}{3} \end{bmatrix}. \tag{10.22}$$

The results of multiplying Eqs. (10.17) and (10.18) by (23) and (12), respectively, from the right are seen to be

$$\frac{1}{6}\mathcal{NPN}(23) = \frac{1}{3}\left[\frac{1}{2}I - \frac{1}{2}(12) - \frac{1}{2}(13) + (23) - (123) + \frac{1}{2}(132)\right],$$

$$\frac{1}{3}\mathcal{NP}(12) = \frac{1}{3}\left[-I + (12) + (123) - (23)\right],$$

and we obtain

$$\frac{1}{3}\mathcal{NP}\left[\frac{1}{2}I - \frac{1}{2}(12)\right] = \frac{1}{6}\mathcal{NPN}, \tag{10.23}$$

$$\frac{1}{3}\mathcal{NP}\left[-\frac{1}{2}I - (12)\right] = \frac{1}{6}\mathcal{NPN}(23). \tag{10.24}$$

For completeness we also give the inverse transformation:

$$\frac{1}{6}\mathcal{NPN}\left[\frac{4}{3}I - \frac{2}{3}(23)\right] = \frac{1}{3}\mathcal{NP}, \tag{10.25}$$

$$\frac{1}{6}\mathcal{NPN}\left[-\frac{2}{3}I - \frac{2}{3}(23)\right] = \frac{1}{3}\mathcal{NP}(12). \tag{10.26}$$

We now return to the problem, and, using the first row of the matrix in Eq. (10.21), we see that

$$\begin{bmatrix} 2p_1 & 2p_2 \\ 2p_3 & \end{bmatrix} = \frac{1}{2}\left(\begin{bmatrix} 2p_1 & 2p_2 \\ 2p_3 & \end{bmatrix}_R - \begin{bmatrix} 2p_3 & 2p_2 \\ 2p_1 & \end{bmatrix}_R\right), \tag{10.27}$$

$$= \frac{1}{2}\left(\begin{bmatrix} 2p_1 & 2p_2 \\ 2p_3 & \end{bmatrix}_R - \begin{bmatrix} 2p_2 & 2p_3 \\ 2p_1 & \end{bmatrix}_R\right). \tag{10.28}$$

This result does not quite finish the problem, however, in that it deals with *unnorma-lized* functions. The coefficients that we show are given assuming the tableau functions of either sort are individually normalized to 1. We must therefore consider some normalization integrals.

The normalization and overlap integrals of the two standard tableaux functions may be written as a matrix

$$S_{ij}^{stf} = \langle 2p_12p_22p_3|\pi_i^{-1}\theta\mathcal{NPN}\pi_j|2p_12p_22p_3\rangle,$$

$$S^{stf} = \begin{bmatrix} 0.365\,144\,70 & 0.182\,572\,36 \\ & 0.313\,656\,66 \end{bmatrix}. \tag{10.29}$$

The corresponding normalization and overlap integrals for the HLSP functions are then obtained with the transformation of Eq. (10.22),

$$S^R = \begin{bmatrix} 0.463\,976\,0 & -0.266\,313\,37 \\ & 0.463\,976\,03 \end{bmatrix}. \tag{10.30}$$

It is seen that the two diagonal elements of $S^R$ are equal, reflecting the symmetrical equivalence of the two Rumer tableaux and diagrams. The coefficients in the wave functions Eqs. (10.15) and (10.16) are all appropriate for each individual tableau function's being normalized to 1. Therefore, $(1/\sqrt{S_{11}^{stf}})\theta \mathcal{N} \mathcal{P} \mathcal{N} 2p_1 2p_2 2p_3$ is a normalized standard tableaux function, with a similar expression for the HLSP functions. In terms of normalized tableau functions we have

$$\frac{1}{\sqrt{S_{11}^{stf}}} \begin{bmatrix} 2p_1 & 2p_2 \\ 2p_3 \end{bmatrix}^u = \frac{1}{2}\sqrt{\frac{S_{11}^R}{S_{11}^{stf}}} \left( \frac{1}{\sqrt{S_{11}^R}} \begin{bmatrix} 2p_1 & 2p_2 \\ 2p_3 \end{bmatrix}_R^u \right.$$

$$\left. - \frac{1}{\sqrt{S_{22}^R}} \begin{bmatrix} 2p_2 & 2p_3 \\ 2p_1 \end{bmatrix}_R^u \right), \tag{10.31}$$

where we have designated unnormalized tableau functions with a superscript "$u$". We now see that $\frac{1}{2}\sqrt{S_{11}^R/S_{11}^{stf}}$ should convert the coefficient of the standard tableaux function in Eq. (10.15) to the coefficient of the HLSP function in Eq. (10.16), i.e.,

$$0.730\,79 \times \frac{1}{2}\sqrt{\frac{0.463\,976\,0}{0.365\,144\,70}} = 0.411\,88. \tag{10.32}$$

For a system of any size, these considerations are tedious and best done with a computer.

### 10.1.3 SCVB treatment with corresponding orbitals

The SCVB method can also be used to study the $\pi$ system of the allyl radical. As we have seen already, only one of the two standard tableaux functions is required because of the symmetry of the molecule. We show the results in Table 10.4, where we see that one arrives at 85% of the correlation energy from the largest MCVB calculation in Table 10.2. There is no entry in Table 10.4 for the EGSO weight, since it would be 1, of course.

The single standard tableaux function is

$$\begin{bmatrix} 2p_1 & 2p_2 \\ 2p_3 \end{bmatrix},$$

Table 10.4. *Results of SCVB calculation.*

| | |
|---|---|
| −116.433 248 63 au | SCF energy |
| −116.470 933 15 au | SCVB energy |
| 1.025 eV | Correlation energy |

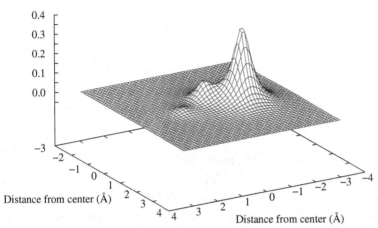

Orbital amplitude

Figure 10.1. The first SCVB orbital for the allyl radical. The orbital amplitude is given in a plane parallel to the radical and 0.5 Å distant.

and the orbitals satisfy

$$\sigma_{yz}2p_1 = 2p_3, \tag{10.33}$$

$$\sigma_{yz}2p_2 = 2p_2, \tag{10.34}$$

$$\sigma_{yz}2p_3 = 2p_1, \tag{10.35}$$

each one consisting of a linear combination of all of the $\pi$ AOs allowed by symmetry. In terms of HLSP functions the wave function has two terms, of course:

$$0.537\,602\,87\left(\begin{bmatrix}2p_2 & 2p_1 \\ 2p_3 \end{bmatrix}_R - \begin{bmatrix}2p_2 & 2p_3 \\ 2p_1 \end{bmatrix}_R\right),$$

and the overlap between the two HLSP functions is −0.730 003.

In Fig. 10.1 we show an altitude drawing of the orbital amplitude of the first of the SCVB orbitals of the allyl $\pi$ system. The third can be obtained by merely reflecting this one in the $y$–$z$ plane of the molecule. It is seen to be concentrated at

Orbital amplitude

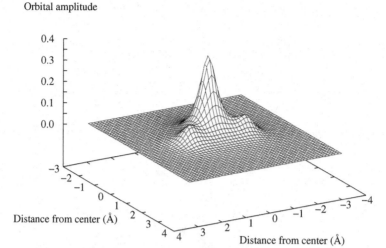

Figure 10.2. The second SCVB orbital for the allyl radical. The orbital amplitude is given in a plane parallel to the radical and 0.5 Å distant.

one end of the radical, with the amplitude falling off fairly rapidly as one moves away from that end. The second SCVB orbital is shown in Fig. 10.2. It is seen to be concentrated on the middle carbon atom with lobes symmetrically placed on either end carbon. Both of these drawings are plotted for amplitudes in a plane 0.5 Å from the plane in which the nuclei occur. Since these are $\pi$ orbitals, the amplitude is, of course, zero in the nuclear plane.

## 10.2 The $He_2^+$ ion

The $He_2^+$ ion has the archetype three-electron bond originally described by Pauling [1], and this section gives a description of MCVB calculation and SCVB treatments for this system. All of these use a Huzinaga 6-G $1s$ function split (411), a 4-G $2s$ function and a $p_z$ function with the scale set to 0.9605. We take up the MCVB treatment first.

### 10.2.1 MCVB calculation

The basis described was used to generate one $1s$ occupied and four virtual RHF orbitals. Using these a full calculation yields 250 standard tableaux functions, which may be combined into 125 functions of $^2\Sigma_u^+$ symmetry. The results for energy, bond distance, and vibrational frequency are shown in Table 10.5. We see that the agreement for $D_e$ is within 0.1 eV, for $R_e$ is within 0.01 Å, and for $\omega_e$ is within $20\,cm^{-1}$. Even at the equilibrium nuclear separation, the wave function is dominated

Table 10.5. *Dissociation energy, bond distance, and vibrational frequency from MCVB calculation of He$_2^+$.*

|       | $D_e$ eV | $R_e$ Å | $\omega_e$ cm$^{-1}$ |
|-------|----------|---------|----------------------|
| Calc. | 2.268    | 1.088 8 | 1715.8               |
| Exp.  | 2.365    | 1.080 8 | 1698.5               |

Table 10.6. *Energy differences $E_{SCVB} - E_{MCVB}$ for He$_2^+$.*

| $\Delta E(R_{min})$ eV | $\Delta E(R_\infty)$ eV |
|------------------------|-------------------------|
| 1.088                  | 1.214                   |

by the first term, and only the second is of further importance,

$$\Psi\left(^2\Sigma_u^+\right) = 0.967\,975 \left(\begin{bmatrix} 1s_a & 1s_a \\ 1s_b & \end{bmatrix} - \begin{bmatrix} 1s_b & 1s_b \\ 1s_a & \end{bmatrix}\right)$$
$$- 0.135\,988 \left(\begin{bmatrix} 2s_a & 2s_a \\ 1s_b & \end{bmatrix} - \begin{bmatrix} 2s_b & 2s_b \\ 1s_a & \end{bmatrix}\right) + \cdots . \quad (10.36)$$

### 10.2.2 SCVB with corresponding orbitals

The three orbitals we use are two we label $1s_a$ and $1s_b$ that are symmetrically equivalent and one $2p_\sigma$ that has the symmetry indicated. Thus if $\sigma_h$ is the horizontal reflection from $D_{\infty h}$ we have the transformations

$$\sigma_h 1s_a = 1s_b,$$
$$\sigma_h 1s_b = 1s_a,$$
$$\sigma_h 2p_\sigma = -2p_\sigma.$$

When these orbitals are optimized, the energies of the SCVB wave functions are higher, of course, than those of the full MCVB wave functions. We show the differences at the equilibrium and infinite internuclear separations in Table 10.6. The energy curves are parallel within $\approx 0.1$ eV, but the SCVB energy is about 1.1 eV higher.

Because of the spatial symmetry there is only one configuration (as with allyl), and in this case the HLSP function function is the simpler of the two. We have for

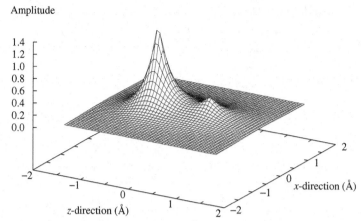

Figure 10.3. The first SCVB orbital for the $He_2^+$ ion. The orbital amplitude is given in the $x$–$z$ plane, which contains the nuclei.

the two forms

$$\Psi_{\text{SCVB}} = \begin{bmatrix} 1s_a & 1s_b \\ 2p_\sigma & \end{bmatrix}_R ,$$

$$= 1.282\,557\,82 \begin{bmatrix} 1s_a & 1s_b \\ 2p_\sigma & \end{bmatrix} - 0.803\,090\,63 \begin{bmatrix} 1s_a & 2p_\sigma \\ 1s_b & \end{bmatrix}, \quad (10.37)$$

where each of the tableaux functions is individually normalized. The second standard tableaux function on the right hand side of Eq. (10.37) is of pure $^2\Sigma_g^+$ symmetry, as can be seen by methods we have used above. Thus the other tableau is of mixed symmetry, and the second term subtracts out the "wrong" part from the first.[2]

The $1s_a$ orbital is shown in Fig. 10.3, and it is seen to be located predominately on one of the nuclei. We may compare this orbital to that for $H_2$ given in Section 3.2.2. The present one is seen to be more localized near the nuclei, reflecting the larger nuclear charge for He. The $1s_b$ orbital is obtained by reflecting with $\sigma_h$. The $2p_\sigma$ orbital is shown as an altitude drawing in Fig. 10.4, where it is seen to have the symmetry indicated by its symbol.

## 10.3 The valence orbitals of the BeH molecule

In this section we give the results of MCVB and SCVB treatments of BeH using a conventional 6-31G** basis.[3] Although there are some similarities to the $He_2^+$ ion, the lack of $g$–$u$ symmetry in this case introduces a number of interesting

---

[2] The relative values of the coefficients in Eq. (10.37) are not determined by the variation theorem, but are imposed by the symmetry and overlaps.

[3] That is, there is a set of $d$ orbitals on Be and a set of $p$ orbitals on H.

Table 10.7. *Dissociation energy, bond distance,*
*vibrational frequency, and electric dipole moment*
*from full MCVB calculation of BeH.*

|  | $D_e$ eV | $R_e$ Å | $\omega_e$ cm$^{-1}$ | $\mu$ D |
|---|---|---|---|---|
| Calc. | 1.963 | 1.348 6 | 211 7.1 | 0.262 |
| Exp. | 2.034 | 1.342 6 | 206 0.8 | |

Amplitude

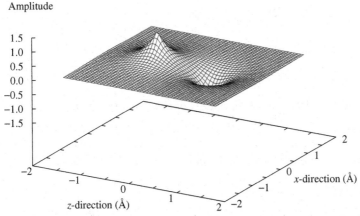

Figure 10.4. The third SCVB orbital for the He$_2^+$ ion. The orbital amplitude is given in the $x$–$z$ plane.

differences. The use of corresponding orbitals does not arise, since there is no symmetry requirement to be satisfied.

### 10.3.1 Full MCVB treatment

A full MCVB calculation on BeH with the above basis yields 504 doublet standard tableaux functions, and these combine into 344 $^2\Sigma^+$ symmetry functions. In Table 10.7 we give some details of the results with experimental values for comparison. The calculated $D_e$ is within 0.1 eV of the experimental value, the values of $R_e$ are quite close, and the vibrational frequency is within 60 cm$^{-1}$. An experimental value for the dipole moment is apparently not known.

The principal configurations in the wave function are shown as HLSP functions in Table 10.8 and as standard tableaux functions in Table 10.9. Considering the HLSP functions, the first is the ground state configuration of the separated atoms, the next two are bonding functions with the $s$–$p$ hybrid of Be and the fourth contributes polarization to the Be$2p_z$ component. The corresponding entries in the third and fourth columns of Table 10.9 do not include the tableau function with the $3p_z$ orbital,

Table 10.8. *The principal HLSP function configurations in BeH at $R = R_e$.*

| Coef. | Tableau |
|---|---|
| 0.332 76 | $\begin{bmatrix} 2s & 2s \\ 1s & \end{bmatrix}_R$ |
| 0.295 79 | $\begin{bmatrix} 1s & 2p_z \\ 2s & \end{bmatrix}_R$ |
| −0.280 28 | $\begin{bmatrix} 2s & 1s \\ 2p_z & \end{bmatrix}_R$ |
| 0.126 09 | $\begin{bmatrix} 1s & 3p_z \\ 2s & \end{bmatrix}_R$ |

Table 10.9. *Large components of the wave function for distances on either side of the cross-over from negative to positive dipole moments. The $n = 1$ orbitals are on H and the $n = 2$ or 3 orbitals are on Be. The Be1s orbitals from the core are omitted in the tableaux.*

| $R = 1.0$ Å | | $R = R_e$ | |
|---|---|---|---|
| Coef. | Tableau | Coef. | Tableau |
| 0.552 01 | $\begin{bmatrix} 2s & 1s \\ 2p_z & \end{bmatrix}$ | 0.530 40 | $\begin{bmatrix} 2s & 1s \\ 2p_z & \end{bmatrix}$ |
| −0.272 34 | $\begin{bmatrix} 2s & 2s \\ 1s & \end{bmatrix}$ | −0.332 76 | $\begin{bmatrix} 2s & 2s \\ 1s & \end{bmatrix}$ |
| 0.187 00 | $\begin{bmatrix} 2s & 1s \\ 3p_z & \end{bmatrix}$ | 0.152 74 | $\begin{bmatrix} 2s & 1s \\ 3p_z & \end{bmatrix}$ |
| 0.107 65 | $\begin{bmatrix} 2p_z & 2p_z \\ 1s & \end{bmatrix}$ | 0.117 30 | $\begin{bmatrix} 1s & 1s \\ 2s & \end{bmatrix}$ |
| −0.079 44 | $\begin{bmatrix} 1s & 3p_z \\ 2p_z & \end{bmatrix}$ | 0.103 57 | $\begin{bmatrix} 2p_z & 2p_z \\ 1s & \end{bmatrix}$ |
| 0.059 29 | $\begin{bmatrix} 1s & 1s \\ 2p_z & \end{bmatrix}$ | 0.089 61 | $\begin{bmatrix} 1s & 1s \\ 2p_z & \end{bmatrix}$ |

but do include the ionic functions involving $1s^2 2s$. Overall, therefore the two different sorts of functions give similar pictures of the bonding in this system.

We now consider the dipole moment, which we will analyze in terms of standard tableaux functions only. Our calculated value for the moment is fairly small, but it is

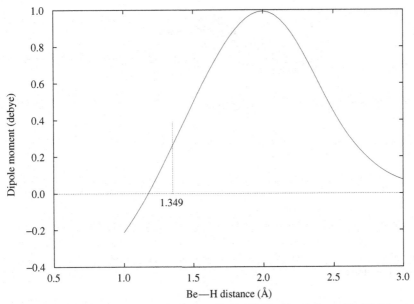

Figure 10.5. The dipole moment function from the MCVB calculation of BeH. The vertical dotted line marks the calculated equilibrium internuclear distance.

in the "right" direction according to the electronegativity difference. The molecule is oriented with the Be atom in the positive $z$-direction in these calculations. We show the dipole as a function of distance in Fig. 10.5. One sees that the $R_e$ of BeH is fairly close to a point at which the moment changes sign to the "wrong" direction. In Table 10.9 we give the major tableau in the wave function on either side of the crossover to see how the dipole moment depends upon distance. At both distances the first three tableaux are covalent and do not have large moments. We saw, however, in Chapter 8 that covalent functions could have small moments in the direction of less diffuse orbitals. The biggest difference here is the overlap of the H1$s$ with the Be1$s$, and we expect the covalent functions to have small negative moments. The H1$s$ and Be2$s$ orbitals are not so different in size and will not contribute so much. The fourth function at $R_e$ is ionic with a large moment in the positive direction. This sort of function does not come in until the sixth place at $R = 1.0$ Å and then with a coefficient only half the size. Thus the main contribution to a positive moment recedes as the distance gets smaller.

### 10.3.2 An SCVB treatment

The allyl radical and the $He_2^+$ ion both have end-for-end symmetry and thus the corresponding orbital SCVB treatment is applied. Consequently, there was only one tableau function in each of those cases. BeH is different in this regard. In the

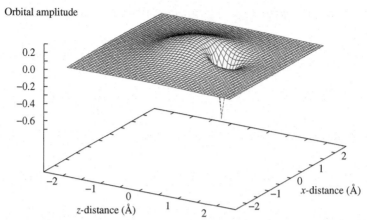

Figure 10.6. The first SCVB orbital for the BeH molecule and associated with the Be nucleus. This has the general appearance of an $s-p$ hybrid pointed toward the H atom, and we denote it the *inner* hybrid, $h_i$. The orbital amplitude is given in the $x-z$ plane, which contains the nuclei.

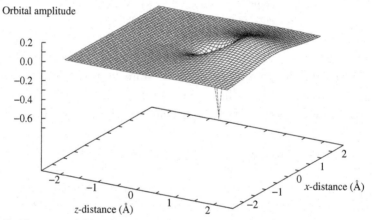

Figure 10.7. The second SCVB orbital for the BeH molecule and associated with the Be nucleus. This has the general appearance of an $s-p$ hybrid pointed away from the H atom, and we denote it the *outer* hybrid, $h_o$. The orbital amplitude is given in the $x-z$ plane, which contains the nuclei.

wave function there are three different orbitals and, consequently, two independent tableaux (of either sort) and an extra variation parameter associated with their mixing. To a considerable extent we may associate two of the orbitals with the Be nucleus and one with the H nucleus. Altitude drawings of the three orbitals are shown in Figs. 10.6, 10.7, and 10.8. They are all orthogonal to the Be$1s$ core orbitals and this results in the sharp negative peak at the Be nucleus. The two Be orbitals in Figs. 10.6 and 10.7 have the general characters of $s-p$ hybrids

Table 10.10. *Energy differences between*
*SCVB and MCVB treatment of BeH.*

| $\Delta E(R_e)$ eV | $\Delta E(R_\infty)$ eV |
|---|---|
| 0.653 | 0.772 |

Orbital amplitude

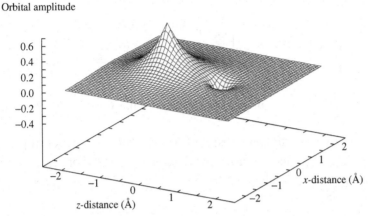

Figure 10.8. The SCVB orbital for the BeH molecule associated with the H nucleus. The orbital amplitude is given in the $x$–$z$ plane, which contains the nuclei.

pointed toward and away from the H atom, respectively. We denote these the inner and outer hybrids, $h_i$ and $h_o$. The third orbital shown in Fig. 10.8 and located mainly on the H atom we will denote simply by $1s$. The energy differences at $R_e$ and $\infty$ geometries are in Table 10.10, where it is seen that the two curves are parallel within about 0.12 eV.

The extra variation parameter with the two tableaux that occur here is shown by the coefficients in Table 10.11, where we use the orbital symbols defined above. The representation of the total wave function is rather similar with the two different sorts of tableau functions. Nevertheless, the HLSP functions have a slight edge in that the perfect pairing function between the inner hybrid and the H$1s$ is a better single-function approximation to the wave function than any of the other tableaux. This is very clear from the EGSO weights that are given.

## 10.4 The Li atom

As we stated in Chapter 4, the Li atom has a much deeper and narrower potential for three electrons than does the allyl radical. One consequence is that the nuclear attraction part of the dynamical effects is relatively more important in Li. Because

Table 10.11. *Coefficients and tableaux for standard tableaux functions and HLSP functions for SCVB treatment of BeH.*
*The orbital symbols are defined in the text.*

| Standard tableaux functions | | | HLSP functions | | |
|---|---|---|---|---|---|
| Coef. | EGSO Wt | Tableau | Coef. | EGSO Wt | Tableau |
| 1.003 58 | 0.877 5 | $\begin{bmatrix} h_i & 1s \\ h_o \end{bmatrix}$ | 1.009 08 | 0.9991 | $\begin{bmatrix} h_i & 1s \\ h_o \end{bmatrix}_R$ |
| −0.235 80 | 0.122 5 | $\begin{bmatrix} h_i & h_o \\ 1s \end{bmatrix}$ | 0.011 28 | 0.0009 | $\begin{bmatrix} 1s & h_o \\ h_i \end{bmatrix}_R$ |

it resembles the SCF result, we, in this case, take up the SCVB wave function first.

These Li atom calculations used Huzinaga's (10/73) basis set[48], further split to (10/73/5221) to yield four basis functions. This is an "*s*" only basis, so our treatments will not produce any angular correlation, but the principles are well illustrated, nevertheless.

### *10.4.1 SCVB treatment*

There is no added symmetry in this example to cause one of the standard tableaux functions to disappear. Thus, the SCVB wave function is

$$\Psi = A \begin{bmatrix} 1s & 1s' \\ 2s \end{bmatrix} + B \begin{bmatrix} 1s & 2s \\ 1s' \end{bmatrix}, \qquad (10.38)$$

where $1s$, $1s'$, and $2s$ are three different linear combinations of the four basis functions. In this case the tableaux in Eq. (10.38) can be interpreted as either the standard or the Rumer sort. The energies and wave functions obtained are shown in Table 10.12. We observe that the wave function in terms of HLSP functions is a little simpler in that the function with $1s$ and $1s'$ coupled to singlet is very nearly all of it. It has been observed that correlation energies are frequently close to 1 eV per pair of electrons, particularly in atoms. The value in Table 10.12 is only a third of that. This is to be expected since we have included the possibility of only radial correlation in our wave function.[4]

---

[4] We do not go into this, but only observe that there are three directions in which electrons may avoid one another. In many cases each direction contributes approximately $1/3$ of the correlation energy.

Table 10.12. *Results of SCVB calculation. The A and B*
*in line 4 are defined in Eq. (10.38).*

| | | |
|---|---|---|
| −7.432 300 22 au | SCF energy | |
| −7.444 280 860 8 au | SCVB energy | |
| 0.326 eV | Correlation energy | |
| | *A* | *B* |
| Standard tableaux | 1.012 989 45 | −0.156 041 34 |
| Rumer tableaux | 0.993 746 95 | −0.006 921 54 |

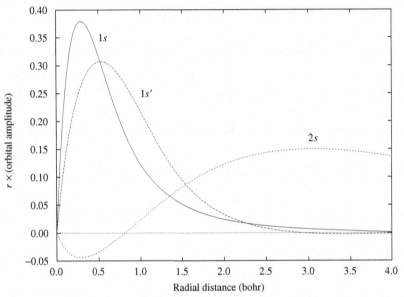

Figure 10.9. The three SCVB orbitals for the Li atom. The orbital amplitudes times the radial coordinate are shown.

In terms of the primitive split Gaussian basis, we obtain for the three SCVB orbitals

$$1s = 0.670\,361\,39s_1 + 0.421\,010\,56s_2 + 0.094\,469\,35s_3 - 0.094\,700\,67s_4,$$
$$1s' = 0.196\,626\,76s_1 + 0.861\,563\,17s_2 - 0.088\,741\,86s_3 + 0.089\,095\,19s_4,$$
$$2s = -0.071\,164\,88s_1 - 0.102\,382\,30s_2 + 0.683\,095\,99s_3 + 0.365\,043\,02s_4,$$

where $s_1, \ldots, s_4$ are the four functions in increasing order of diffuseness. These SCVB orbitals are plotted in Fig. 10.9 where we see that the inner shell orbitals are split and show radial correlation. The outer orbital has one radial node like the familiar $2s$ orbital of SCF theory, but in this case it is exactly orthogonal to neither of the inner orbitals.

Table 10.13. *Results of MCVB calculation for Li.*

| | |
|---|---|
| $-7.432\,300\,22$ au | SCF energy |
| $-7.446\,822\,77$ au | MCVB energy |
| $0.395$ eV | Correlation energy |
| $0.9746$ | EGSO pop. of $\begin{bmatrix} 1s & 1s' \\ 2s & \end{bmatrix}$ |

### 10.4.2 MCVB treatment

We now describe the full MCVB treatment of the Li atom using a basis consisting of the three SCVB orbitals $1s$, $1s'$, and $2s$ to which we add the primitive $s_4$ for completeness. The Hamiltonian matrix is $20 \times 20$, and the energy is of course the same regardless of the sort of CI performed, so long as it is "full". The results are shown in Table 10.13, where we see that the SCVB calculation arrived at about 83% of the correlation energy available from this basis. The EGSO population of the principal SCVB standard tableaux function is very high. The additional correlation energy from the MCVB is principally from intershell correlation and is produced by the accumulation of a number of configurations with fairly small coefficients.

# 11

## Second row homonuclear diatomics

For many years chemists have considered that an understanding of the theory of the bonding of the homonuclear diatomic molecules from the second row of the periodic table is central to understanding all of bonding, and we consider these stable molecules first from our VB point of view. The stable molecules with interesting multiple bonds are $B_2$, $C_2$, $N_2$, $O_2$, and $F_2$. Of course, $F_2$ has only a single bond by ordinary bonding rules, but we include it in our discussion. $Li_2$ is stable, but, qualitatively, is similar to $H_2$. The question of the existence of $Be_2$ is also interesting, but is really a different sort of problem from that of the other molecules. Of the five molecules we do consider, $B_2$ and $C_2$ are known only spectroscopically, while the other three exist at room temperature all around us or in the laboratory.

### 11.1 Atomic properties

Before we launch into the discussion of the molecules, we examine the nature of the atoms we are dealing with. As we should expect, this has a profound effect on the structure of the molecules we obtain. We show in Fig. 11.1 the first few energy levels of B through F with the ground state taken at zero energy. The $L$-$S$ term symbols are also shown. The ground configurations of B and F each support just one term, $^2P$, but the other three support three terms. All of these are at energies below $\approx 4.2$ eV (relative to their ground state energies). The states from the ground configuration are connected with lines marked with a G in Fig. 11.1.

We also consider two sorts of excited states.

- Those arising from configurations that differ from the ground configuration by one $2s \rightarrow 2p$ transition. This may be called a *valence* excited configuration. The lines connecting these states are marked with a "V".
- Those arising from a $2p \rightarrow 3s$ transition. This is the first configuration of a Rydberg-like series. The lines connecting these states are marked with an "R".

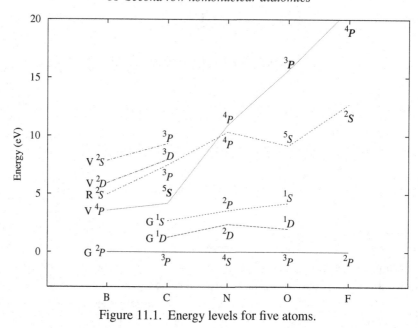

Figure 11.1. Energy levels for five atoms.

In the cases of B and C three terms have been observed from the valence excited configuration. Apparently only one has been observed for the others. Only one state has been plotted for the Rydberg state. We note that the energy of the lowest valence excited state crosses the lowest Rydberg state energy between C and N. As two identical atoms from the group approach one another to form a molecule, the ground and several excited states mix together in the final structure. The general point we make is that we expect valence states to be more important than Rydberg states in bonding,[1] and the atoms at the start of the list have much lower valence state energies than those later, and the valence states should mix in more easily upon molecule formation. The energy necessary for the excitation of the atomic states must come from the bonding energy. We analyze the bonding of these molecules using the above general ideas.

## 11.2 Arrangement of bases and quantitative results

A goal of this chapter is to show, for the diatomic molecules under discussion, both the capability of the VB method in providing quantitative estimates of molecular properties and its capability of giving qualitative pictures of the bonding. The quantitative results are illustrated in Table 11.1, where we give values for $R_e$, the equilibrium bond distance, and $D_e$,[2] determined theoretically with STO3G, 6-31G,

---

[1] Among other things, the $3s$ orbital is too diffuse to overlap other orbitals well.
[2] $D_0 + \omega_e/2$, see Huber and Herzberg[49].

Table 11.1. *Bond distances and $D_e$ for some second row homonuclear diatomics.*

| Source | | $B_2$ | $C_2$ | $N_2$ | $O_2$ | $F_2$ |
|--------|--------|-------|-------|-------|-------|-------|
| Exp. | $R_e$ Å | 1.590 | 1.243 | 1.098 | 1.208 | 1.412 |
| | $D_e$ eV | 3.085 | 6.325 | 9.905 | 5.214 | 1.659 |
| STO3G | $R_e$ | 1.541 | 1.261 | 1.198 | 1.263 | 1.392 |
| | $D_e$ | 3.778 | 6.851 | 6.452 | 4.529 | 2.082 |
| 6-31G | $R_e$ | 1.623 | 1.243 | 1.137 | 1.290 | 1.502 |
| | $D_e$ | 2.781 | 6.164 | 7.666 | 3.830 | 1.565 |
| 6-31G* | $R_e$ | 1.610 | 1.261 | 1.129 | 1.257 | 1.436 |
| | $D_e$ | 3.053 | 6.331 | 8.768 | 4.580 | 1.867 |

Table 11.2. *Number of symmetry functions for 6-31G and 6-31G* basis sets.*

| Basis | $B_2$ | $C_2$ | $N_2$ | $O_2$ | $F_2$ |
|-------|-------|-------|-------|-------|-------|
| 6-31G | 777 | 991 | 1086 | 697 | 109 |
| 6-31G* | 1268 | 1864 | 1812 | 1188 | 131 |

and 6-31G* bases compared with those from experiment. All of the VB calculations here have the two $1s^2$ shells in each atom completely closed, and, in order to save space, none of the tableaux shown in tables later in this chapter include these orbitals explicitly.

In Chapter 9 we gave a somewhat generic description of the way we arrange basis sets. More details are given here.

**STO3G** Each of the treatments may be described as a *full valence* calculation. The details of numbers of VB structures is given below in Table 11.4.

**6-31G** The VB structure basis is a full valance set augmented by structures involving a single excitation from one valence orbital to one virtual orbital, using all possible combinations of the excitation (outside the $1s$ shells). Table 11.2 shows the number of symmetry functions (the dimension of the $H$ and $S$ matrices) for each case.

**6-31G*** The AO basis in this case is the same as that for the 6-31G set with a set of $d$ orbitals added. In these calculations the $d_\sigma$ and the $d_\pi$ orbitals are included in the virtual orbital set in which single excitations are included in generating structures. The $d_\delta$ orbitals were not used. The inclusion of these $d$ orbitals provides polarization when the molecule is formed.

An examination of the values in Table 11.1 shows a variety of results for the different molecules and bases. In general, the values of $R_e$ are too large except for

$B_2$ and $F_2$ values using the STO3G basis. The 6-31G* values average about 0.03 Å too large.

The calculated $D_e$ values vary somewhat more randomly. In general the values for $B_2$ and $C_2$ are the closest to the experimental ones, followed by those for $F_2$. The values of $D_e$ for $N_2$ are the hardest to obtain followed by those for $O_2$. With the 6-31G* basis set, the calculated $D_e$ values for $B_2$, $C_2$, and $F_2$ are all within 0.2 eV of experiment, while $N_2$ is off by more than 1 eV. It is not clear why $N_2$ presents such a challenge.

## 11.3 Qualitative discussion

If one wishes a qualitative picture of the bonding and structure of a molecule it has become evident that this is most easily determined from a reasonably minimal basis calculation. As one increases the size of the basis, the set of important structures remains reasonably stable, but there is frequently some jockeying around. As we argued earlier, the STO3G set was historically optimized to be appropriate for molecular geometries, therefore it is, perhaps, not surprising that it gives a reasonable picture of molecular structure, even when taken over to the VB method. In spite of this bias toward the molecular state, the STO3G basis also gives a good account of the states that the system migrates into as the separation between the atoms goes to $\infty$. In the present section we therefore examine the wave functions obtained with this basis for the molecules we are discussing to determine the VB picture of their bonding.

In Table 11.3 we show the ground states of the atoms and the ground state of the diatomic molecules they form. Except possibly for $B_2$, all of these are well established spectroscopically. This same table shows the total degeneracy for two infinitely separated atoms. For example, atomic boron is in a $^2P^o$ state, which, ignoring spin-orbit coupling (i.e., using the ESE), is six-fold degenerate. Each of these states can couple with each in another atom, so, all together, we expect $6 \times 6 = 36$ different states to come together at $\infty$. This will include singlets, triplets, $\Sigma$, $\Pi$, and $\Delta$ states with various $g$ and $u$ and $+$ and $-$ labels, but the number will add up to 36. We are not interested in discussing most of these but the interested reader can make calculations for each of the symmetries with CRUNCH.

As we stated earlier, all of the configurations we use have the two $1s^2$ shells occupied. Thus if we allow all possible occupations of the remaining eight valence orbitals in the STO3G basis, we may speak of a *full valence* VB. We have this same number of valence orbitals in all of the molecules we treat this way. As we pass from $B_2$ to $F_2$, the number of electrons that the orbitals must hold increases, however, causing a considerable variation in the number of allowed states. We show the

Table 11.3. *Atomic ground states and asymptotic molecular symmetries.*

| Atom | Atomic state | Molecular states Deg. at $\infty$ | Bonding state |
|------|-------------|------------------|---------------|
| B | $^2P^o$ | 36 | $^3\Sigma_g^-$ |
| C | $^3P^e$ | 81 | $^1\Sigma_g^+$ |
| N | $^4S^o$ | 16 | $^1\Sigma_g^+$ |
| O | $^3P^e$ | 81 | $^3\Sigma_g^-$ |
| F | $^2P^o$ | 36 | $^1\Sigma_g^+$ |

Table 11.4. *Statistics of full valence calculations for STO3G basis.*

| Molecule | $2S+1$ | Number of electrons[a] | Number of constellations | Number of symmetry functions |
|----------|--------|------------------------|--------------------------|------------------------------|
| $B_2(^3\Sigma_g^-)$ | 3 | 6 | 18 | 41 |
| $C_2(^1\Sigma_g^+)$ | 1 | 8 | 35 | 84 |
| $N_2(^1\Sigma_g^+)$ | 1 | 10 | 76 | 102 |
| $O_2(^3\Sigma_g^-)$ | 3 | 12 | 18 | 30 |
| $F_2(^1\Sigma_g^+)$ | 1 | 14 | 8 | 8 |

[a] Outside a $1s_a^2 1s_b^2$ core.

numbers in Table 11.4, where it is seen that the total size of the variational problem is a maximum for the case of $N_2$. We now proceed to a discussion of the individual molecules.

### 11.3.1 $B_2$

The very first question that we might ask is: From our knowledge of the properties of VB functions and knowing that the atom is in a $^2P^o$ state, can we predict the likely ground state symmetry of the molecule? With $B_2$ this may be tricky. We list some conjectures.

1. The very first guess might be that, outside of the two $2s^2$ closed subshells, a single $\sigma$ bond is formed from the two $p$ orbitals in the $\sigma$ orientation. A singlet state is expected.
2. A more intricate situation arises if the excited configuration, $2s2p^2$, can come into play. Then the two $2s$ and two $2p_\sigma$ electrons can each form an electron pair bond, but there are still two $2p_\pi$ electrons hanging around.

Table 11.5. *Principal standard tableaux function structures for $B_2$ at an asymptotic bond distance.*

|  | 1 | 2 | 3 |
|---|---|---|---|
| Num.[a] | 2 | 8 | 8 |
| Tab.[b] | $\begin{bmatrix} 2s_a & 2s_a \\ 2s_b & 2s_b \\ p_{xa} \\ p_{yb} \end{bmatrix}$ | $\begin{bmatrix} 2s_a & 2s_a \\ p_{xb} & p_{xb} \\ p_{xa} \\ p_{yb} \end{bmatrix}$ | $\begin{bmatrix} p_{xa} & p_{xa} \\ p_{yb} & p_{yb} \\ p_{xb} \\ p_{ya} \end{bmatrix}$ |
| $C_i(\infty)$ | 0.665 111 24 | −0.118 177 21 | 0.020 997 77 |

[a] The number of terms in the symmetry function that is generated from the tableau shown. (See text.)
[b] These tableau symbols exclude the core orbitals.

Table 11.6. *Principal HLSP function structures for $B_2$ at an asymptotic bond distance.*

|  | 1 | 2 | 3 |
|---|---|---|---|
| Num.[a] | 2 | 8 | 8 |
| Tab.[b] | $\begin{bmatrix} 2s_a & 2s_a \\ 2s_b & 2s_b \\ p_{xa} \\ p_{yb} \end{bmatrix}_R$ | $\begin{bmatrix} 2s_a & 2s_a \\ p_{xb} & p_{xb} \\ p_{xa} \\ p_{yb} \end{bmatrix}_R$ | $\begin{bmatrix} p_{xa} & p_{xa} \\ p_{yb} & p_{yb} \\ p_{xb} \\ p_{ya} \end{bmatrix}_R$ |
| $C_i(\infty)$ | 0.665 111 24 | −0.118 177 21 | 0.020 997 77 |

[a] The number of terms in the symmetry function that is generated from the tableau shown. (See text.)
[b] These tableau symbols exclude the core orbitals.

(a) The two $2p_\pi$ orbitals could form an electron pair bond. Thus we would expect a $^1\Delta_g$ state with three bonds.
(b) The two $2p_\pi$ electrons could arrange themselves in two one-electron bonds, one for the $x$-direction and one for the $y$-direction.
   i. If the two electrons are singlet coupled, we have a $^1\Sigma_g^+$ state.
   ii. If the two electrons are triplet coupled, we have a $^3\Sigma_g^-$ state.

We have collected some results for standard tableaux functions and HLSP functions in Tables 11.5 and 11.6. The structure of these tables will be repeated in several later sections, and we describe it here.

1. The unlabeled row gives the ordinal number of the following entries.
2. The "Num." row gives the number of tableau functions in a symmetry function. Thus the "2" for column one indicates that the tableau function below it is the first of a two function sum that has the correct $1\Sigma_g^+$ symmetry.
3. The "Tab." row gives the actual tableaux in terms of AO symbols.

Table 11.7. *Principal standard tableaux function structures for* $B_2$
*at energy minimum bond distance.*

|  | 1 | 2 | 3 | 4 |
|---|---|---|---|---|
| Num. | 2 | 2 | 2 | 2 |
| Tab. | $\begin{bmatrix} 2s_a & 2s_b \\ p_{za} & p_{zb} \\ p_{xa} \\ p_{yb} \end{bmatrix}$ | $\begin{bmatrix} 2s_a & 2s_a \\ 2s_b & p_{za} \\ p_{xb} \\ p_{yb} \end{bmatrix}$ | $\begin{bmatrix} 2s_a & 2s_a \\ 2s_b & 2s_b \\ p_{xa} \\ p_{yb} \end{bmatrix}$ | $\begin{bmatrix} 2s_a & 2s_b \\ p_{za} & p_{yb} \\ p_{zb} \\ p_{xa} \end{bmatrix}$ |
| $C_i(R_{\min})$ | 0.215 367 05 | −0.206 418 77 | 0.198 518 19 | 0.115 676 19 |

4. The $C_i$ are the coefficients in the wave function corresponding to the tableaux. The values assume that the actual tableau function is normalized to 1 as well as the overall wave function. The "$\infty$" indicates the values in this case are for large $R$-values. Elsewhere, other arguments appear.

We return to consideration of the entries in these tables, where we give the principal structures for two B atoms at long distance. It can be seen that we did not need to give two tables since they are the same. The reader should recall that the two sorts of VB functions are the same when there is only one standard tableau, as is the case here. Focusing on Table 11.5 we see that the principal structure involves the orbitals of the atomic configuration $2s^2 2p$ on each atom. The relatively small coefficient is caused by the fact that the principal structure is really

$$0.665\,111\,24 \left( \begin{bmatrix} 2s_a & 2s_a \\ 2s_b & 2s_b \\ p_{xa} \\ p_{yb} \end{bmatrix} + \begin{bmatrix} 2s_a & 2s_a \\ 2s_b & 2s_b \\ p_{xb} \\ p_{ya} \end{bmatrix} \right),$$

i.e., there are two terms in the symmetry function. If the symmetry function were normalized in the form $(f + g)/\sqrt{2}$ the coefficient would be $\approx 0.941$. We also emphasize that the footnotes in Tables 11.5 and 11.6 apply equally well to all of the tables in this chapter that show tableaux and coefficients.

The second and third terms involve excited states that produce electron correlation (particularly of the angular sort) in the closed $2s^2$ shells of the atoms. Therefore, the wave function for the asymptotic geometry is essentially the product of two atomic wave functions.

When the two atoms are at the geometry of the energy minimum the results are as shown in Tables 11.7 and 11.8, where, as before, we give both the standard tableaux function and HLSP function results. It is clear that the wave function is now a more complicated mixture of many structures. In addition, the apparent importance of the structures based upon the values of the coefficients is somewhat different for the

Table 11.8. *Principal HLSP function structures for* $B_2$ *at the energy minimum bond distance.*

| | 1 | 2 | 3 | 4 |
|---|---|---|---|---|
| Num. | 2 | 2 | 2 | 2 |
| Tab. | $\begin{bmatrix} 2s_a & 2s_a \\ 2s_b & 2s_b \\ p_{xa} \\ p_{yb} \end{bmatrix}_R$ | $\begin{bmatrix} 2s_a & 2s_a \\ 2s_b & p_{za} \\ p_{xb} \\ p_{yb} \end{bmatrix}_R$ | $\begin{bmatrix} 2s_b & p_{za} \\ 2s_a & p_{zb} \\ p_{xa} \\ p_{yb} \end{bmatrix}_R$ | $\begin{bmatrix} 2s_a & 2s_a \\ 2s_b & 2s_b \\ p_{xa} \\ p_{ya} \end{bmatrix}_R$ |
| $C_i(R_{\min})$ | 0.198 518 19 | −0.134 559 43 | −0.118 190 06 | 0.097 154 09 |

standard tableaux function and HLSP function bases. Considering Table 11.7 first, we see that structure 1 consists of orbitals from the excited valence configuration of each atom, $2s2p^2$. Structure 4 is another of the nine standard tableaux from this arrangement. Structure 2 has the orbitals of one atom in the ground state and one in the excited valence state. Structure 3 is from the two atoms each in their ground states. Thus, the VB picture of the $B_2$ molecule consists of roughly equal parts of these three atomic configurations. There are, of course, many smaller terms leading to electron correlation.

The picture from the results of Table 11.8 is not significantly different. Structure 3 from before is now 1 (with the same coefficient, of course), but we have a mixture of the same atomic configurations. The new structures 2 and 3 show standard two-electron bonds involving $2s$ and $2p_z$ orbitals on opposite atoms. This feature is not so clear from the standard tableaux functions.

### 11.3.2 $C_2$

The ground state of the C atom is $^3P$ from a $2s^2 2p^2$ configuration. In the case of B we saw that the excited valence configuration played an important role in the structures describing the $B_2$ molecule. $C_2$ has more electrons with the possibility of more bonds, and, thus, there may be more tendency for the valence excited configuration to be important in this molecule than in $B_2$.

Our conjectures concerning the lowest state of $C_2$ are as follows.

1. The two $^3P$ atoms could form a $p_\sigma$–$p_\sigma$ bond with the two remaining $p_\pi$ orbitals coupled:
   (a) $^3\Sigma^-$ as in $B_2$;
   (b) $^1\Pi$ to give a doubly degenerate ground state.
2. The two $^3P$ atoms could form two $p_\pi$ bonds to produce a $^1\Sigma^+$ ground state. If the valence excited state is important as argued above, these two $^5S$ states could also couple to $^1\Sigma^+$ interacting strongly with the two $p_\pi$ bonds.

It is this last situation that pertains.

Table 11.9. *Principal standard tableaux function structures*
*for $C_2$ at an asymptotic bond distance.*

|  | 1 | 2 | 3 |
|---|---|---|---|
| Num. | 2 | 4 | 2 |
| Tab. | $\begin{bmatrix} 2s_a & 2s_a \\ 2s_b & 2s_b \\ p_{za} & p_{zb} \\ p_{xa} & p_{xb} \end{bmatrix}$ | $\begin{bmatrix} 2s_a & 2s_a \\ p_{yb} & p_{yb} \\ p_{za} & p_{zb} \\ p_{xa} & p_{xb} \end{bmatrix}$ | $\begin{bmatrix} p_{xa} & p_{xa} \\ p_{xb} & p_{xb} \\ p_{za} & p_{zb} \\ p_{ya} & p_{yb} \end{bmatrix}$ |
| $C_i(\infty)$ | 0.690 603 71 | $-0.106\,757\,10$ | 0.016 503 07 |

Table 11.10. *Principal HLSP function structures for $C_2$*
*at an asymptotic bond distance.*

|  | 1 | 2 | 3 | 4 |
|---|---|---|---|---|
| Num. | 2 | 2 | 8 | 4 |
| Tab. | $\begin{bmatrix} 2s_a & 2s_a \\ 2s_b & 2s_b \\ p_{za} & p_{zb} \\ p_{xa} & p_{xb} \end{bmatrix}_R$ | $\begin{bmatrix} 2s_a & 2s_a \\ 2s_b & 2s_b \\ p_{zb} & p_{xa} \\ p_{za} & p_{xb} \end{bmatrix}_R$ | $\begin{bmatrix} 2s_a & 2s_a \\ p_{yb} & p_{yb} \\ p_{za} & p_{zb} \\ p_{xa} & p_{xb} \end{bmatrix}_R$ | $\begin{bmatrix} 2s_a & 2s_a \\ p_{yb} & p_{yb} \\ p_{za} & p_{zb} \\ p_{xa} & p_{xb} \end{bmatrix}_R$ |
| $C_i(\infty)$ | 0.398 720 24 | $-0.398\,720\,24$ | $-0.061\,636\,24$ | 0.009 528 05 |

First, however, we examine the asymptotic geometry. The principal structures are shown in Tables 11.9 and 11.10. In the standard tableaux function case structure 1 is one of the possible $^1\Sigma^+$ couplings of two $^3P$ atoms, and structures 2 and 3 produce electron correlation in the closed $2s$ shell. The results with HLSP functions are essentially the same with some differences in the coefficients. The apparently smaller coefficients in the latter case result mainly from the larger number of terms in the symmetry functions.

We show the standard tableaux function results for the energy minimum geometry in Table 11.11. Here we see that the valence excited configuration has become the dominant structure, and the $^1\Sigma_g^+$ coupling of the $^3P$ atomic ground states is structure 2. Structure 3 is a mixture of the $^5S$ and $^3P$ states while structure 4 is another of the standard tableau associated with structure 1.

We call attention to a significant similarity between structure 1 and the tableau for the $^5S$ atomic state, which is

$$\begin{bmatrix} 2s \\ 2p_z \\ 2p_x \\ 2p_y \end{bmatrix}.$$

Table 11.11. *Principal standard tableaux function structures for $C_2$ at the energy minimum bond distance.*

| | 1 | 2 | 3 | 4 |
|---|---|---|---|---|
| Num. | 1 | 1 | 4 | 1 |
| Tab. | $\begin{bmatrix} 2s_a & 2s_b \\ p_{za} & p_{zb} \\ p_{xa} & p_{xb} \\ p_{ya} & p_{yb} \end{bmatrix}$ | $\begin{bmatrix} 2s_a & 2s_a \\ 2s_b & 2s_b \\ p_{xa} & p_{xb} \\ p_{ya} & p_{yb} \end{bmatrix}$ | $\begin{bmatrix} 2s_b & 2s_b \\ p_{ya} & p_{ya} \\ 2s_a & p_{zb} \\ p_{xa} & p_{xb} \end{bmatrix}$ | $\begin{bmatrix} 2s_a & 2s_b \\ p_{za} & p_{xb} \\ p_{zb} & p_{ya} \\ p_{xa} & p_{yb} \end{bmatrix}$ |
| $C_i(R_{min})$ | 0.438 636 13 | 0.293 039 91 | 0.158 969 69 | −0.131 203 05 |

Table 11.12. *Principal HLSP function structures for $C_2$ at the energy minimum bond distance.*

| | 1 | 2 | 3 | 4 |
|---|---|---|---|---|
| Num. | 1 | 1 | 2 | 2 |
| Tab. | $\begin{bmatrix} 2s_a & 2s_a \\ 2s_b & 2s_b \\ p_{xa} & p_{xb} \\ p_{ya} & p_{yb} \end{bmatrix}_R$ | $\begin{bmatrix} 2s_b & p_{za} \\ 2s_a & p_{zb} \\ p_{xa} & p_{xb} \\ p_{ya} & p_{yb} \end{bmatrix}_R$ | $\begin{bmatrix} 2s_a & 2s_a \\ p_{yb} & p_{yb} \\ 2s_b & p_{za} \\ p_{xa} & p_{xb} \end{bmatrix}_R$ | $\begin{bmatrix} 2s_b & 2s_b \\ p_{ya} & p_{ya} \\ 2s_a & p_{zb} \\ p_{xa} & p_{xb} \end{bmatrix}_R$ |
| $C_i(R_{min})$ | 0.240 701 83 | −0.179 769 81 | −0.165 030 55 | 0.127 710 28 |

This function is antisymmetric with respect to the interchange of any pair of orbitals. The same pertains to structure 1 of Table 11.11 with respect to either of the columns. Thus the dominant structure is very much two $^5S$ atoms.

The results for HLSP functions in Table 11.12 show a somewhat different picture. In this case the dominant (but not by much) structure is the one with two $\pi$ bonds and structures 3 and 4 provide a $\sigma$ bond. Structure 2 is the double $^5S$ structure, but, since HLSP functions do not have a close relationship to the actual $^5S$ state as above, there is less importance to just one Rumer coupling scheme.

### 11.3.3 $N_2$

We commented above that the energies of the first excited valence states of B and C are fairly low and there is a large jump between C and N. The reason for this is principally the Coulomb repulsion energy in the states. For B and C the excited valence state has one less paired orbital than the corresponding ground state, while for N, O, and F the numbers are the same. Since the Coulomb repulsion energy tends to be largest between two electrons in the same orbital, this trend is not surprising.

Table 11.13. *Principal standard tableaux function structure for $N_2$ at an asymptotic bond distance.*

| | 1 |
|---|---|
| Num. | 1 |
| Tab. | $\begin{bmatrix} 2s_a & 2s_a \\ 2s_b & 2s_b \\ p_{za} & p_{zb} \\ p_{xa} & p_{xb} \\ p_{ya} & p_{yb} \end{bmatrix}$ |
| $C_i(\infty)$ | 1.000 001 54 |

Table 11.14. *Principal HLSP function structures for $N_2$ at an asymptotic bond distance.*

| | 1 | 2 | 3 |
|---|---|---|---|
| Num. | 1 | 1 | 1 |
| Tab. | $\begin{bmatrix} 2s_a & 2s_a \\ 2s_b & 2s_b \\ p_{zb} & p_{xa} \\ p_{za} & p_{xb} \\ p_{ya} & p_{yb} \end{bmatrix}_R$ | $\begin{bmatrix} 2s_a & 2s_a \\ 2s_b & 2s_b \\ p_{xa} & p_{xb} \\ p_{zb} & p_{ya} \\ p_{za} & p_{yb} \end{bmatrix}_R$ | $\begin{bmatrix} 2s_a & 2s_a \\ 2s_b & 2s_b \\ p_{za} & p_{zb} \\ p_{xb} & p_{ya} \\ p_{xa} & p_{yb} \end{bmatrix}_R$ |
| $C_i(\infty)$ | 0.471 403 79 | 0.471 403 79 | 0.471 403 79 |

In addition, we are comparing these molecules with a minimal basis. With eight valence orbitals and ten electrons, configurations that produce some angular correlation in the $2s$ shell cannot occur in the asymptotic region. The upshot is that there is just one principal standard tableaux function at long distance, and this is shown in Table 11.13. Because of the antisymmetry in the columns of standard tableaux functions, we see that this function represents two noninteracting $^4S$ N atoms.

The situation is not so simple with HLSP functions. They do not have the antisymmetry characteristic mentioned above, and the asymptotic state requires a sum of three of them as shown in Table 11.14.

When two $^4S$ N atoms form a molecule we have the possibility that there could be three bonds, one from the two $p_\sigma$ orbitals, and two from the four $p_\pi$ orbitals. Some mixing of the $2s$ with the $p_\sigma$ orbitals might lead to hybridization. No other possibilities seem likely. We show the principal configurations in the HLSP function and standard tableaux function cases in Tables 11.15 and 11.16, respectively. We see that the same orbitals are present in both main structures. The situation with

Table 11.15. *Principal HLSP function structures for $N_2$ at the energy minimum bond distance.*

| | 1 | 2 | 3 | 4 |
|---|---|---|---|---|
| Num. | 1 | 1 | 2 | 4 |
| Tab. | $\begin{bmatrix} 2s_a & 2s_a \\ 2s_b & 2s_b \\ p_{za} & p_{zb} \\ p_{xa} & p_{xb} \\ p_{ya} & p_{yb} \end{bmatrix}_R$ | $\begin{bmatrix} 2s_a & 2s_a \\ p_{zb} & p_{zb} \\ 2s_b & p_{za} \\ p_{xa} & p_{xb} \\ p_{ya} & p_{yb} \end{bmatrix}_R$ | $\begin{bmatrix} 2s_a & 2s_a \\ 2s_b & 2s_b \\ p_{xa} & p_{xa} \\ p_{yb} & p_{yb} \\ p_{za} & p_{zb} \end{bmatrix}_R$ | $\begin{bmatrix} 2s_a & 2s_a \\ 2s_b & 2s_b \\ p_{zb} & p_{zb} \\ p_{ya} & p_{ya} \\ p_{xa} & p_{xb} \end{bmatrix}_R$ |
| $C_i(R_{\min})$ | 0.207 439 81 | 0.103 862 35 | 0.081 907 16 | −0.075 261 88 |

Table 11.16. *Principal standard tableaux function structures for $N_2$ at the energy minimum bond distance.*

| | 1 | 2 | 3 | 4 | 5 |
|---|---|---|---|---|---|
| Num. | 1 | 1 | 2 | 1 | 1 |
| Tab. | $\begin{bmatrix} 2s_a & 2s_a \\ 2s_b & 2s_b \\ p_{za} & p_{zb} \\ p_{xa} & p_{xb} \\ p_{ya} & p_{yb} \end{bmatrix}$ | $\begin{bmatrix} 2s_a & 2s_a \\ 2s_b & 2s_b \\ p_{za} & p_{xb} \\ p_{zb} & p_{ya} \\ p_{xa} & p_{yb} \end{bmatrix}$ | $\begin{bmatrix} 2s_a & 2s_a \\ p_{zb} & p_{zb} \\ 2s_b & p_{za} \\ p_{xa} & p_{xb} \\ p_{ya} & p_{yb} \end{bmatrix}$ | $\begin{bmatrix} 2s_a & 2s_a \\ 2s_b & 2s_b \\ p_{za} & p_{xa} \\ p_{zb} & p_{xb} \\ p_{ya} & p_{yb} \end{bmatrix}$ | $\begin{bmatrix} 2s_a & 2s_a \\ 2s_b & 2s_b \\ p_{za} & p_{zb} \\ p_{xa} & p_{ya} \\ p_{xb} & p_{yb} \end{bmatrix}$ |
| $C_i(R_{\min})$ | 0.329 868 28 | −0.158 776 96 | 0.112 116 70 | −0.111 073 50 | −0.110 745 84 |

the HLSP functions is somewhat simpler. The main structure has three electron pair bonds involving the $2p$ orbitals, and structure 2 involves one atom in the first excited valence state with an electron pair bond between $2s$ and $2p_\sigma$ orbitals. This latter occurrence, of course, indicates a certain amount of $s-p$ hybridization in the $\sigma$ bond. Structures 3 and 4 represent ionic contributions to the $\pi$ and $\sigma$ bonds, respectively.

The results for the standard tableaux functions at the energy minimum are shown in Table 11.16. Structures 1, 2, 4, and 5 are different standard tableaux corresponding to two ground state atoms and represent mixing in different states from the ground configurations. The standard tableaux functions are not so simple here since they do not represent three electron pair bonds as a single tableau. Structure 3 represents one of the atoms in the first excited valence state and contributes to $s-p$ hybridization in the $\sigma$ bond as in the HLSP function case.

It is clear that, regardless of the sort of basis function we use, our results give the bonding picture of $N_2$ as a triple bond. There is, in addition, some indication that the excited valence configuration is less important compared to the ground configuration than was the case with $B_2$ and $C_2$. Two properties of the atoms could contribute to this.

- As already mentioned, the excited valence state is of higher energy and is less likely to mix as strongly.
- Exciting the atom in this case does not change the number of paired electrons, and, thus, a no greater opportunity for bonding presents itself than in the ground state.

### 11.3.4 $O_2$

From Table 11.3 we see that the ground state of O is $^3P$, and there are only two unpaired orbitals in the ground configuration. Since the $L$ shell is more than half full, valence excitations will not reduce the number of double occupations. We can make the following conjectures.

1. The two free $2p$ orbitals from each atom could combine to form two $\pi$ bonds to give a $^1\Sigma_g^+$ state.
2. One of the $2p$ orbitals on each atom could join with the other to form a $\sigma$ bond.
   (a) The other $2p$ orbitals could combine as a pair of $\pi$ bonds to give a $^1\Pi_u$ state.
   (b) The other single $2p$ orbitals could combine with the doubly occupied orbital on the other atom to form two *three-electron bonds*[1], giving a $^3\Sigma_g^-$ state.

It is, of course, the last case that occurs, and we consider first the nature of a three-electron bond.

Any elementary inorganic structure book will describe, in MO terms, the $\pi$ bonds in $O_2$ as each having a doubly occupied bonding orbital and a singly occupied antibonding orbital. (This is the MO description of a three-electron bond.) We may analyze this description, using the properties of tableau functions, to see how it relates to the VB picture.

We take a very simple case of a pair of orbitals $a$ and $b$ that can bond. We assume the orbitals are at two different centers. The simplest LCAO approximation to the bonding orbital is $\sigma = A(a + b)$, and the antibonding counterpart is $\sigma^* = B(a - b)$. Here $A = 1/\sqrt{2(1 + S)}$ and $B = 1/\sqrt{2(1 - S)}$, where $S$ is the overlap integral, are the normalization constants. Consider the simple three-electron doublet wave function

$$\psi = \begin{bmatrix} \sigma & \sigma \\ \sigma^* & \end{bmatrix}, \tag{11.1}$$

$$= A^2 B \begin{bmatrix} a + b & a + b \\ a - b & \end{bmatrix}. \tag{11.2}$$

Table 11.17. *Principal standard tableaux*
*function structures for $O_2$ at an asymptotic*
*bond distance.*

|       | 1 | 2 |
|-------|---|---|
| Num.  | 2 | 2 |
| Tab.  | $\begin{bmatrix} 2s_a & 2s_a \\ 2s_b & 2s_b \\ 2p_{xa} & 2p_{xa} \\ 2p_{yb} & 2p_{yb} \\ 2p_{za} & 2p_{zb} \\ 2p_{xb} & \\ 2p_{ya} & \end{bmatrix}$ | $\begin{bmatrix} 2s_a & 2s_a \\ 2s_b & 2s_b \\ 2p_{xa} & 2p_{xa} \\ 2p_{yb} & 2p_{yb} \\ 2p_{za} & 2p_{xb} \\ 2p_{zb} & \\ 2p_{ya} & \end{bmatrix}$ |
| $C_i(\infty)$ | 0.612 374 06 | $-0.612\,374\,06$ |

It will be recalled from our discussion of Chapter 5 that the tableau in Eq. (11.2) is a shorthand for the result of operating upon a particular orbital product with the operator $\theta\,\mathcal{NPN}$, and $\mathcal{N}$ is the column antisymmetrizer. Thus, our function contains a $2 \times 2$ functional determinant involving $a \pm b$ and two particles in all terms. Any row or column operations legal in a determinant may be used to simplify our function, and the determinant may be converted to the equal one involving just $2a$ and $-b$. Equation (11.2) becomes

$$\psi = A^2 B \begin{bmatrix} 2a & a+b \\ -b & \end{bmatrix}, \tag{11.3}$$

$$= 2A^2 B \left( \begin{bmatrix} b & b \\ a & \end{bmatrix} - \begin{bmatrix} a & a \\ b & \end{bmatrix} \right), \tag{11.4}$$

and shows us how the three-electron bond is represented in the VB scheme. We also emphasize that the tableaux of Eqs. (11.2) and (11.4) are of the sort where the standard tableaux functions and the HLSP functions are the same. Thus, that distinction does not affect our picture.

Now, let us consider the principal structures for the asymptotic geometry shown in Tables 11.17 and 11.18. Both forms of the wave function correspond to $^3\Sigma_g^-$ couplings of the two atoms in their $^3P$ ground states.

When we consider the principal structures at the energy minimum geometry we see the three-electron bonds discussed above. These are shown in Tables 11.19 and 11.20. Considering the principal tableaux of either sort, we see there are two three-electron sets present, $p_{xa}^2 p_{xb}$ and $p_{yb}^2 p_{ya}$. There is, of course, a normal two-electron $\sigma$ bond present also. When we move to the second structure, there are differences.

Table 11.18. *Principal HLSP function structures*
*for $O_2$ at an asymptotic bond distance.*

|  | 1 | 2 |
|---|---|---|
| Num. | 2 | 2 |
| Tab. | $\begin{bmatrix} 2s_a & 2s_a \\ 2s_b & 2s_b \\ 2p_{xa} & 2p_{xa} \\ 2p_{yb} & 2p_{yb} \\ 2p_{za} & 2p_{zb} \\ 2p_{xb} & \\ 2p_{ya} & \end{bmatrix}_R$ | $\begin{bmatrix} 2s_a & 2s_a \\ 2s_b & 2s_b \\ 2p_{xa} & 2p_{xa} \\ 2p_{yb} & 2p_{yb} \\ 2p_{ya} & 2p_{xb} \\ 2p_{zb} & \\ 2p_{za} & \end{bmatrix}_R$ |
| $C_i(\infty)$ | 0.500 000 0 | 0.500 000 0 |

Table 11.19. *Principal standard tableaux function structures for $O_2$ at the energy*
*minimum bond distance.*

|  | 1 | 2 | 3 | 4 | 5 |
|---|---|---|---|---|---|
| Num. | 2 | 2 | 2 | 2 | 2 |
| Tab. | $\begin{bmatrix} 2s_a & 2s_a \\ 2s_b & 2s_b \\ 2p_{xa} & 2p_{xa} \\ 2p_{yb} & 2p_{yb} \\ 2p_{za} & 2p_{zb} \\ 2p_{xb} & \\ 2p_{ya} & \end{bmatrix}$ | $\begin{bmatrix} 2s_a & 2s_a \\ 2s_b & 2s_b \\ 2p_{xa} & 2p_{xa} \\ 2p_{ya} & 2p_{ya} \\ 2p_{za} & 2p_{zb} \\ 2p_{xb} & \\ 2p_{yb} & \end{bmatrix}$ | $\begin{bmatrix} 2s_a & 2s_a \\ 2s_b & 2s_b \\ 2p_{xa} & 2p_{xa} \\ 2p_{yb} & 2p_{yb} \\ 2p_{za} & 2p_{xb} \\ 2p_{zb} & \\ 2p_{ya} & \end{bmatrix}$ | $\begin{bmatrix} 2s_a & 2s_a \\ 2s_b & 2s_b \\ 2p_{za} & 2p_{za} \\ 2p_{xb} & 2p_{xb} \\ 2p_{yb} & 2p_{yb} \\ 2p_{xa} & \\ 2p_{ya} & \end{bmatrix}$ | $\begin{bmatrix} 2s_a & 2s_a \\ 2s_b & 2s_b \\ 2p_{xb} & 2p_{xb} \\ 2p_{ya} & 2p_{ya} \\ 2p_{za} & 2p_{xa} \\ 2p_{zb} & \\ 2p_{yb} & \end{bmatrix}$ |
| $C_i(R_{\min})$ | 0.385 676 56 | −0.190 603 83 | −0.189 268 76 | 0.175 541 11 | −0.164 445 74 |

Table 11.20. *Principal HLSP function structures for $O_2$ at the energy*
*minimum bond distance.*

|  | 1 | 2 | 3 | 4 |
|---|---|---|---|---|
| Num. | 2 | 2 | 2 | 4 |
| Tab. | $\begin{bmatrix} 2s_a & 2s_a \\ 2s_b & 2s_b \\ 2p_{xa} & 2p_{xa} \\ 2p_{yb} & 2p_{yb} \\ 2p_{za} & 2p_{zb} \\ 2p_{xb} & \\ 2p_{ya} & \end{bmatrix}_R$ | $\begin{bmatrix} 2s_a & 2s_a \\ 2s_b & 2s_b \\ 2p_{za} & 2p_{za} \\ 2p_{xb} & 2p_{xb} \\ 2p_{yb} & 2p_{yb} \\ 2p_{xa} & \\ 2p_{ya} & \end{bmatrix}_R$ | $\begin{bmatrix} 2s_a & 2s_a \\ 2s_b & 2s_b \\ 2p_{xa} & 2p_{xa} \\ 2p_{ya} & 2p_{ya} \\ 2p_{za} & 2p_{zb} \\ 2p_{xb} & \\ 2p_{yb} & \end{bmatrix}_R$ | $\begin{bmatrix} 2s_a & 2s_a \\ 2s_b & 2s_b \\ 2p_{za} & 2p_{za} \\ 2p_{xa} & 2p_{xa} \\ 2p_{yb} & 2p_{yb} \\ 2p_{xb} & \\ 2p_{ya} & \end{bmatrix}_R$ |
| $C_i(R_{\min})$ | 0.327 172 36 | 0.175 541 11 | −0.171 075 55 | −0.113 001 07 |

Table 11.21. *Principal standard tableaux*
*and HLSP function structures for* $F_2$
*at an asymptotic bond distance.*

|  | 1 |
| --- | --- |
| Num. | 2 |
| Tab. | $\begin{bmatrix} 2s_a & 2s_a \\ 2s_b & 2s_b \\ p_{za} & p_{za} \\ p_{zb} & p_{zb} \\ p_{xa} & p_{xa} \\ p_{xb} & p_{xb} \\ p_{ya} & p_{yb} \end{bmatrix}$ |
| $C_i(\infty)$ | 0.707 106 78 |

1. With standard tableaux functions:
   (a) structure 2 is ionic, having the two three-electron bonds pointed the same way;
   (b) structures 3 and 4 are the other standard tableau associated with structure 1;
   (c) structure 5 makes ionic contributions to all bonds, but in such a way that the net charge on the atoms is zero. The charge in a three-electron bond is one way, and the charge in the $\sigma$ bond is opposite.
2. With HLSP functions:
   (a) structure 2 is ionic with a zero net atomic charge. This is similar to structure 5 in terms of the standard tableaux functions;
   (b) structure 3 is ionic with a net charge. The two three-electron bonds point in the same direction;
   (c) structure 4 is ionic with respect to the $\sigma$ bond.

### 11.3.5 $F_2$

As we pass to $F_2$, with a minimal basis the amount of flexibility remaining is small. The only unpaired orbital in the atom is a $2p$ one, and these are expected to form a $\sigma$ electron pair bond and a $^1\Sigma_g^+$ molecular state. In fact, with 14 electrons and 8 orbitals (outside the core) there can be, at most, one unpaired orbital set in any structure. Therefore, in this case there is no distinction between the standard tableaux and HLSP function representations of the wave functions, and we give only one set of tables. As is seen from Table 11.21, there is only one configuration present at asymptotic distances. That shown is one of the $^1\Sigma_g^+$ combinations of two $^2P$ atoms.

Table 11.22 shows the principal structures at the energy minimum bond distance. Structure 1 is a $\sigma$ bond comprising the two $p_\sigma$ orbitals, and structure 2 is

Table 11.22. *Principal standard tableaux and HLSP function structures for $F_2$ at the energy minimum bond distance.*

|  | 1 | 2 | 3 | 4 |
|---|---|---|---|---|
| Num. | 1 | 2 | 2 | 2 |
| Tab. | $\begin{bmatrix} 2s_a & 2s_a \\ 2s_b & 2s_b \\ p_{xa} & p_{xa} \\ p_{xb} & p_{xb} \\ p_{ya} & p_{ya} \\ p_{yb} & p_{yb} \\ p_{za} & p_{zb} \end{bmatrix}$ | $\begin{bmatrix} 2s_a & 2s_a \\ 2s_b & 2s_b \\ p_{zb} & p_{zb} \\ p_{xa} & p_{xa} \\ p_{xb} & p_{xb} \\ p_{ya} & p_{ya} \\ p_{yb} & p_{yb} \end{bmatrix}$ | $\begin{bmatrix} 2s_a & 2s_a \\ p_{zb} & p_{zb} \\ p_{xa} & p_{xa} \\ p_{xb} & p_{xb} \\ p_{ya} & p_{ya} \\ p_{yb} & p_{yb} \\ 2s_b & p_{za} \end{bmatrix}$ | $\begin{bmatrix} 2s_a & 2s_a \\ p_{za} & p_{za} \\ p_{xa} & p_{xa} \\ p_{xb} & p_{xb} \\ p_{ya} & p_{ya} \\ p_{yb} & p_{yb} \\ 2s_b & p_{zb} \end{bmatrix}$ |
| $C_i(R_{min})$ | 0.779 221 33 | −0.232 134 50 | 0.053 534 27 | 0.044 702 64 |

ionic, contributing to correlation in the bond. Structures 3 and 4 contribute to $s$–$p$ hybridization in the bond.

## 11.4 General conclusions

In Section 11.1 we pointed out that B and C atoms have relatively low-lying valence excited states compared to the other atoms considered. It is seen that these valence excited states comprise the principal structures in the bonded state of $B_2$ and $C_2$, but not in the other molecules where they contribute less than the ground configuration. We shall discuss these effects in further detail for C atoms in Chapter 13. If we treat the one- and three-electron bonds as one-half a bond we see that $B_2$, $C_2$, $N_2$, $O_2$, and $F_2$ have two, three, three, two, and one bond(s) in the molecule, respectively. Were it not for the low-lying valence excited states in B and C, the molecules corresponding to these might be expected to have one and two bonds, respectively. Nevertheless, the more open structure of the valence excited states allows more bonding between the atoms.

The two molecules that have one- or three-electron $\pi$ bonds show triplet ground states. This conforms to Hund's rule in atoms where one has unpaired electrons distributed among degenerate orbitals to produce the highest possible multiplicity. The other molecules all have electron pair bonds or unshared pairs and are in singlet states.

# 12

# Second row heteronuclear diatomics

The consideration of isoelectronic sequences can provide considerable physical understanding of structural details. We here give details of the calculation of a series of isoelectronic diatomic molecules from the second row of the periodic table, $N_2$, CO, BF, and BeNe. By studying this sequence we see how the competition between nuclear charges affects bonding. All of these are closed-shell singlet systems, and, at least in the cases of the first two, conventional bonding arguments say there is a triple bond between the two atoms. We expect, at most, only a Van der Waals type of bond between Be and Ne, of course. Our calculations should predict this.

The three polar molecules in the series are interesting because they all have anomalous directions to their dipole moments, i.e., the direction is different from that predicted by an elementary application of the idea of electronegativity, accepting the fact that there may be ambiguity in the definition of electronegativity for Ne. We will see how VB ideas interpret these anomalous dipole moments.

We do the calculations with a 6-31G* basis in the same way as was done in Chapter 11 and for three arrangements of STO3G bases. This will allow us both to judge the stability of the qualitative predictions to the basis and to assess the ability of the calculations to obtain quantitative answers.

We have already treated $N_2$ in Chapter 11, but will look at it here from a somewhat different point of view.

## 12.1 An STO3G AO basis

Results of calculations carried out with three different selection schemes and an STO3G AO will be described. The reader will recall that the scale factors for this basis are traditionally adjusted to give molecular geometries, and this must be remembered when interpreting the results. By now the reader should suspect that such a basis will not produce very accurate energies. Nevertheless, we see that the qualitative trends of the quantities match the experimental values.

Table 12.1. *Dissociation energies and equilibrium distances for isoelectronic series with an STO3G basis and a full calculation. Energies are in electron volts and distances are in ångstroms.*

| Basis | | $N_2$ | CO | BF | BeNe |
|---|---|---|---|---|---|
| | Exp. | | | | |
| | $D_e$ | 9.905 | 11.226 | 7.897 | |
| | $R_m$ | 1.098 | 1.128 | 1.263 | |
| STO3G | ($\pm 1$, grouped) | | | | |
| | $D_e$ | 6.101 | 8.988 | 6.917 | 0.023 |
| | $R_m$ | 1.199 | 1.195 | 1.257 | 2.583 |
| | ($\pm 1$, ungrouped) | | | | |
| | $D_e$ | 6.448 | 9.444 | 7.162 | 0.123 |
| | $R_m$ | 1.198 | 1.196 | 1.265 | 2.159 |
| | (full) | | | | |
| | $E_d$ | 6.452 | 9.460 | 7.181 | 0.125 |
| | $R_m$ | 1.265 | 1.196 | 1.264 | 2.151 |
| 6-31G* | Full valence+$S^a$ | | | | |
| | $D_e$ | 8.768 | 11.053 | 7.709 | 0.053 |
| | $R_m$ | 1.129 | 1.155 | 1.278 | 3.066 |

[a] S is an abbreviation for 'single excitations'.

The three ways in which the structures are selected for the calculation follow, and in all cases the $1s$ orbitals of the atoms are doubly occupied.

- "$\pm 1$, grouped" This indicates that the structures included in the VB calculation are restricted to those in which there is at most only one electron transferred from one atom to the other and in which there are six $\sigma$, two $\pi_x$, and two $\pi_y$ electrons.
- "$\pm 1$, ungrouped" This indicates that the structures included in the VB calculation are restricted to those in which there is at most only one electron transferred from one atom to the other.
- "full" This is the full (valence) VB calculation.

Dissociation energies and minimum energy atomic separations from the STO3G bases are given in Table 12.1 along with those for the 6-31G* basis, which we will discuss later. We note that the restriction to $\pm 1$ ionicities has an effect on the energy of at most 10–20 meV for this basis.

We give tables of the important structures in the full wave function using spherical AOs and using the $s-p$ hybrids, $2s \pm 2p_z$. The energies are, of course, the same for these alternatives, but the apparent importance of the standard tableaux functions or HLSP functions differs. We also discuss EGSO results for the series.

Again we see that the $D_e$ of $N_2$ is the most poorly predicted in this series. We have no clear explanation for this at present.

Table 12.2. $N_2$: *The most important terms in the wave function when spherical AOs are used as determined by the magnitudes of the coefficients. Results for standard tableaux and HLSP functions are given. See text.*

|  | 1 | 2 | 3 | 4 |
|---|---|---|---|---|
| Num. | 1 | 1 | 2 | 1 |
| STF | $\begin{bmatrix} 2s_a & 2s_a \\ 2s_b & 2s_b \\ 2p_{za} & 2p_{zb} \\ 2p_{xa} & 2p_{xb} \\ 2p_{ya} & 2p_{yb} \end{bmatrix}$ | $\begin{bmatrix} 2s_a & 2s_a \\ 2s_b & 2s_b \\ 2p_{za} & 2p_{xb} \\ 2p_{zb} & 2p_{ya} \\ 2p_{xa} & 2p_{yb} \end{bmatrix}$ | $\begin{bmatrix} 2s_b & 2s_b \\ 2p_{za} & 2p_{za} \\ 2s_a & 2p_{zb} \\ 2p_{xa} & 2p_{xb} \\ 2p_{ya} & 2p_{yb} \end{bmatrix}$ | $\begin{bmatrix} 2s_a & 2s_a \\ 2s_b & 2s_b \\ 2p_{za} & 2p_{xa} \\ 2p_{zb} & 2p_{xb} \\ 2p_{ya} & 2p_{yb} \end{bmatrix}$ |
| $C_i(min)$ | 0.329 86 | −0.158 78 | −0.112 12 | −0.111 07 |
| Num. | 1 | 1 | 2 | 1 |
| HLSP | $\begin{bmatrix} 2s_a & 2s_a \\ 2s_b & 2s_b \\ 2p_{za} & 2p_{zb} \\ 2p_{xa} & 2p_{xb} \\ 2p_{ya} & 2p_{yb} \end{bmatrix}_R$ | $\begin{bmatrix} 2s_a & 2s_a \\ 2p_{zb} & 2p_{zb} \\ 2s_b & 2p_{za} \\ 2p_{xa} & 2p_{xb} \\ 2p_{ya} & 2p_{yb} \end{bmatrix}_R$ | $\begin{bmatrix} 2s_a & 2s_a \\ 2s_b & 2s_b \\ 2p_{xa} & 2p_{xa} \\ 2p_{yb} & 2p_{yb} \\ 2p_{za} & 2p_{zb} \end{bmatrix}_R$ | $\begin{bmatrix} 2s_a & 2s_a \\ 2s_b & 2s_b \\ 2p_{za} & 2p_{za} \\ 2p_{yb} & 2p_{yb} \\ 2p_{xa} & 2p_{xb} \end{bmatrix}_R$ |
| $C_i(min)$ | 0.207 44 | 0.103 86 | 0.081 91 | −0.075 26 |

### 12.1.1 $N_2$

In Table 12.2 we show the four most important structures in the wave function as determined by the magnitude of the coefficients for standard tableaux functions and for HLSP functions. Table 12.3 shows the same information for the $\sigma$ AOs formed into $s$–$p$ hybrids. The symbols "$h_{ox}$" or "$h_{ix}$" represent the outward or inward pointing hybrids, respectively. Using the size of the coefficients as a measure of importance we see that the expected structure involving one $\sigma$ and two $\pi$ bonds is the largest in the wave function. It appears that the hybrid orbital arrangement is slightly preferred for standard tableaux functions while the spherical orbital arrangement is slightly preferred for HLSP functions, but the difference is not great. These results suggest that an intermediate rather than one-to-one hybridization might be preferable, but a great difference is not expected. Nevertheless, it is clear that the VB method predicts a triple bond between the two atoms in $N_2$.

The layout of Tables 12.3 and 12.4 is similar to that of Tables 11.5 and 11.6 described in Section 11.3.1. There is, nevertheless, one point concerning the "Num." row that merits further comment. In Chapter 6 we discussed how the symmetric group projections interact with spatial symmetry projections. Functions 1, 2, and 4 are members of one constellation, and the corresponding coefficients may not be entirely independent. There are three linearly independent $^1\Sigma_g^+$ symmetry functions from the five standard tableaux of this configuration. The 1, 2, and 4 coefficients are thus possibly partly independent and partly connected by group theory. In none

Table 12.3. $N_2$: *The most important terms in the wave function when s–p hybrid AOs are used as determined by the magnitudes of the coefficients. Results for standard tableaux and HLSP functions are given.*

|  | 1 | 2 | 3 | 4 |
|---|---|---|---|---|
| Num. | 1 | 1 | 1 | 2 |
| STF | $\begin{bmatrix} h_{oa} & h_{oa} \\ h_{ob} & h_{ob} \\ h_{ia} & h_{ib} \\ 2p_{xa} & 2p_{xb} \\ 2p_{ya} & 2p_{yb} \end{bmatrix}$ | $\begin{bmatrix} h_{oa} & h_{oa} \\ h_{ob} & h_{ob} \\ h_{ia} & 2p_{xb} \\ h_{ib} & 2p_{ya} \\ 2p_{xa} & 2p_{yb} \end{bmatrix}$ | $\begin{bmatrix} h_{oa} & h_{oa} \\ h_{ob} & h_{ob} \\ h_{ia} & h_{ib} \\ 2p_{xa} & 2p_{ya} \\ 2p_{xb} & 2p_{yb} \end{bmatrix}$ | $\begin{bmatrix} h_{ob} & h_{ob} \\ h_{ia} & h_{ia} \\ h_{oa} & h_{ib} \\ 2p_{xa} & 2p_{xb} \\ 2p_{ya} & 2p_{yb} \end{bmatrix}$ |
| $C_i(min)$ | 0.337 43 | −0.144 76 | −0.107 25 | 0.099 41 |
| Num. | 1 | 4 | 1 | 1 |
| HLSP | $\begin{bmatrix} h_{oa} & h_{oa} \\ h_{ob} & h_{ob} \\ h_{ia} & h_{ib} \\ 2p_{xa} & 2p_{xb} \\ 2p_{ya} & 2p_{yb} \end{bmatrix}_R$ | $\begin{bmatrix} h_{oa} & h_{oa} \\ h_{ob} & h_{ob} \\ h_{ia} & h_{ia} \\ 2p_{xb} & 2p_{xb} \\ 2p_{ya} & 2p_{yb} \end{bmatrix}_R$ | $\begin{bmatrix} h_{oa} & h_{oa} \\ h_{ia} & h_{ia} \\ h_{ob} & h_{ib} \\ 2p_{xa} & 2p_{xb} \\ 2p_{ya} & 2p_{yb} \end{bmatrix}_R$ | $\begin{bmatrix} h_{oa} & h_{oa} \\ h_{ob} & h_{ob} \\ 2p_{xb} & 2p_{xb} \\ h_{ia} & h_{ib} \\ 2p_{ya} & 2p_{yb} \end{bmatrix}_R$ |
| $C_i(min)$ | 0.189 12 | −0.089 64 | 0.086 69 | 0.086 31 |

of the tables do we attempt to elucidate this sort of question. It really requires a detailed examination of the output of the symgenn segment of the CRUNCH suite.

Table 12.4 shows spherical and hybrid AO results when subjected to the EGSO weight analysis. Unlike the coefficients, the EGSO analysis for these results shows that an ionic function is the single structure that contains the largest fraction of the wave function in both of these cases. This is a common result in molecules with multiple bonding. We have seen that the ionic structures contribute to delocalization of the electrons (see Chapter 2) and thereby reduce the kinetic energy of the structure. In a complicated symmetry function involving the sum of several terms the mixing of the improved correlation energy of the covalent functions and the improved kinetic energy of the ionic functions can produce a symmetry constellation that has the highest weight. It is thus important to interpret these results correctly.

We may comment that the principal configuration with spherical AOs measured by coefficient in the wave function,

$$\begin{bmatrix} 2s_a & 2s_a \\ 2s_b & 2s_b \\ 2p_{za} & 2p_{zb} \\ 2p_{xa} & 2p_{xb} \\ 2p_{ya} & 2p_{yb} \end{bmatrix},$$

is not present among the first four functions measured by the EGSO weights. In fact its weight is 0.012 69, a little lower than those in the table. The situation is similar

Table 12.4. $N_2$: EGSO weights (standard tableaux functions) for spherical AOs, upper group, and s–p hybrids, lower group. These are weights for whole symmetry functions *rather than individual tableaux*. It should be recalled from Chapter 6 that the detailed forms of symmetry functions are dependent on the particular arrangement of the orbitals in the tableaux and are frequently nonintuitive.

|       | 1 | 2 | 3 | 4 |
|-------|---|---|---|---|
| Num. | 4 | 2 | 6 | 4 |
| STF | $\begin{bmatrix} 2s_a & 2s_a \\ 2s_b & 2s_b \\ 2p_{xa} & 2p_{xa} \\ 2p_{za} & 2p_{zb} \\ 2p_{ya} & 2p_{yb} \end{bmatrix}$ | $\begin{bmatrix} 2s_a & 2s_a \\ 2s_b & 2s_b \\ 2p_{za} & 2p_{za} \\ 2p_{xa} & 2p_{xb} \\ 2p_{ya} & 2p_{yb} \end{bmatrix}$ | $\begin{bmatrix} 2s_a & 2s_a \\ 2p_{zb} & 2p_{zb} \\ 2p_{xa} & 2p_{xa} \\ 2s_b & 2p_{za} \\ 2p_{ya} & 2p_{yb} \end{bmatrix}$ | $\begin{bmatrix} 2s_a & 2s_a \\ 2p_{za} & 2p_{za} \\ 2p_{xa} & 2p_{xa} \\ 2p_{yb} & 2p_{yb} \\ 2s_b & 2p_{zb} \end{bmatrix}$ |
| Wt | 0.558 62 | 0.200 53 | 0.066 58 | 0.040 89 |
| Num. | 4 | 2 | 6 | 2 |
| STF | $\begin{bmatrix} h_{oa} & h_{oa} \\ h_{ob} & h_{ob} \\ 2p_{xa} & 2p_{xa} \\ h_{ia} & h_{ib} \\ 2p_{ya} & 2p_{yb} \end{bmatrix}$ | $\begin{bmatrix} h_{oa} & h_{oa} \\ h_{ob} & h_{ob} \\ h_{ia} & h_{ia} \\ 2p_{xa} & 2p_{xb} \\ 2p_{ya} & 2p_{yb} \end{bmatrix}$ | $\begin{bmatrix} h_{oa} & h_{oa} \\ h_{ia} & h_{ia} \\ 2p_{yb} & 2p_{yb} \\ h_{ob} & h_{ib} \\ 2p_{xa} & 2p_{xb} \end{bmatrix}$ | $\begin{bmatrix} h_{oa} & h_{oa} \\ h_{ob} & h_{ob} \\ 2p_{xa} & 2p_{xa} \\ 2p_{yb} & 2p_{yb} \\ h_{ia} & h_{ib} \end{bmatrix}$ |
| Wt | 0.611 66 | 0.156 66 | 0.057 14 | 0.038 70 |

with the hybrid orbitals. In this case the standard tableaux function

$$\begin{bmatrix} h_{oa} & h_{oa} \\ h_{ob} & h_{ob} \\ h_{ia} & h_{ib} \\ 2p_{xa} & 2p_{xb} \\ 2p_{ya} & 2p_{yb} \end{bmatrix}$$

has an EGSO weight of 0.005 32, rather smaller than the value in the case of spherical functions. The reader should not find these small contributions too unexpected. The ionic structures singled out by the EGSO can be looked at as one-third ionic and two-thirds covalent. When the orthogonalization inherent in the method works, the effect of the purely covalent functions is considerably depressed and is already taken care of by the mixed functions.

### 12.1.2 CO

The set of tables we give for CO follows the pattern given for $N_2$ in the last section. Table 12.5 shows the four most important structures in the wave function of CO

Table 12.5. *CO: The most important terms in the wave function when spherical AOs are used as determined by the magnitudes of the coefficients. Results for standard tableaux and HLSP functions are given.*

| | 1 | 2 | 3 | 4 |
|---|---|---|---|---|
| Num. | 2 | 1 | 1 | 1 |
| STF | $\begin{bmatrix} 2s_a & 2s_a \\ 2s_b & 2s_b \\ 2p_{xb} & 2p_{xb} \\ 2p_{za} & 2p_{zb} \\ 2p_{ya} & 2p_{yb} \end{bmatrix}$ | $\begin{bmatrix} 2s_a & 2s_a \\ 2s_b & 2s_b \\ 2p_{zb} & 2p_{zb} \\ 2p_{xa} & 2p_{xb} \\ 2p_{ya} & 2p_{yb} \end{bmatrix}$ | $\begin{bmatrix} 2s_a & 2s_a \\ 2s_b & 2s_b \\ 2p_{za} & 2p_{zb} \\ 2p_{xa} & 2p_{xb} \\ 2p_{ya} & 2p_{yb} \end{bmatrix}$ | $\begin{bmatrix} 2s_b & 2s_b \\ 2p_{zb} & 2p_{zb} \\ 2s_a & 2p_{za} \\ 2p_{xa} & 2p_{xb} \\ 2p_{ya} & 2p_{yb} \end{bmatrix}$ |
| $C_i(min)$ | 0.205 59 | −0.192 23 | 0.149 09 | −0.120 44 |
| Num. | 2 | 1 | 2 | 1 |
| HLSP | $\begin{bmatrix} 2s_a & 2s_a \\ 2s_b & 2s_b \\ 2p_{xb} & 2p_{xb} \\ 2p_{za} & 2p_{zb} \\ 2p_{ya} & 2p_{yb} \end{bmatrix}_R$ | $\begin{bmatrix} 2s_a & 2s_a \\ 2s_b & 2s_b \\ 2p_{zb} & 2p_{zb} \\ 2p_{xa} & 2p_{xb} \\ 2p_{ya} & 2p_{yb} \end{bmatrix}_R$ | $\begin{bmatrix} 2s_a & 2s_a \\ 2s_b & 2s_b \\ 2p_{zb} & 2p_{zb} \\ 2p_{xb} & 2p_{xb} \\ 2p_{ya} & 2p_{yb} \end{bmatrix}_R$ | $\begin{bmatrix} 2s_a & 2s_a \\ 2s_b & 2s_b \\ 2p_{xb} & 2p_{xb} \\ 2p_{yb} & 2p_{yb} \\ 2p_{za} & 2p_{zb} \end{bmatrix}_R$ |
| $C_i(min)$ | 0.179 15 | −0.157 016 | −0.094 20 | 0.093 19 |

as determined by the magnitude of the coefficients for standard tableaux functions and HLSP functions. Table 12.6 shows the same information for the $\sigma$ AOs formed into $s-p$ hybrids. The symbols "$h_{ox}$" or "$h_{ix}$" are used as before. Using the size of the coefficients as a measure of importance, we see that VB theory predicts CO to have only two covalent bonds between the atoms. We saw in Section 11.1 that C and O are both in $^3P$ ground states, thus elementary considerations suggest that there is one $\sigma$ covalent bond and one $\pi$ covalent bond cylindrically averaged to achieve $^1\Sigma^+$ symmetry. This view, although too simplistic, is different from that often seen where CO is written like $N_2$ with a triple bond. The latter must also be too simplistic, since, if CO had anything close to an evenly shared triple bond, its dipole moment would be large, although in the experimentally correct direction. We will discuss the dipole moments of the polar molecules all together in Section 12.3.

The triple bond structure appears in the third place with spherical AOs and standard tableaux functions, but is not among the first four with HLSP functions. This is actually misleading due to the arbitrary cutoff at four functions in the table. The HLSP function triple bond has a coefficient of 0.091 82, only slightly smaller that function 4 in the table. The appearance of the triple bond structure in this wave function is the quantum mechanical manifestation of the "$\pi$ back-bonding" phenomenon invoked in qualitative arguments concerning bonding. We thereby have a quantitative approach to the concept.

Table 12.6. CO: *The most important terms in the wave function when s–p hybrid AOs are used as determined by the magnitudes of the coefficients. Results for standard tableaux and HLSP functions are given.*

|  | 1 | 2 | 3 | 4 |
|---|---|---|---|---|
| Num. | 2 | 1 | 1 | 1 |
| STF | $\begin{bmatrix} h_{oa} & h_{oa} \\ h_{ob} & h_{ob} \\ 2p_{yb} & 2p_{yb} \\ h_{ia} & h_{ib} \\ 2p_{xa} & 2p_{xb} \end{bmatrix}$ | $\begin{bmatrix} h_{oa} & h_{oa} \\ h_{ob} & h_{ob} \\ h_{ib} & h_{ib} \\ 2p_{xa} & 2p_{xb} \\ 2p_{ya} & 2p_{yb} \end{bmatrix}$ | $\begin{bmatrix} h_{oa} & h_{oa} \\ h_{ob} & h_{ob} \\ h_{ia} & h_{ib} \\ 2p_{xa} & 2p_{xb} \\ 2p_{ya} & 2p_{yb} \end{bmatrix}$ | $\begin{bmatrix} h_{ob} & h_{ob} \\ h_{ib} & h_{ib} \\ h_{oa} & h_{ia} \\ 2p_{xa} & 2p_{xb} \\ 2p_{ya} & 2p_{yb} \end{bmatrix}$ |
| $C_i(min)$ | 0.274 04 | 0.222 04 | 0.188 09 | 0.146 23 |
| Num. | 2 | 1 | 1 | 2 |
| HLSP | $\begin{bmatrix} h_{oa} & h_{oa} \\ h_{ob} & h_{ob} \\ 2p_{xb} & 2p_{xb} \\ h_{ia} & h_{ib} \\ 2p_{ya} & 2p_{yb} \end{bmatrix}_R$ | $\begin{bmatrix} h_{oa} & h_{oa} \\ h_{ob} & h_{ob} \\ h_{ib} & h_{ib} \\ 2p_{xa} & 2p_{xb} \\ 2p_{ya} & 2p_{yb} \end{bmatrix}_R$ | $\begin{bmatrix} h_{oa} & h_{oa} \\ h_{ob} & h_{ob} \\ 2p_{xb} & 2p_{xb} \\ 2p_{yb} & 2p_{yb} \\ h_{ia} & h_{ib} \end{bmatrix}_R$ | $\begin{bmatrix} h_{oa} & h_{oa} \\ h_{ob} & h_{ob} \\ h_{ib} & h_{ib} \\ 2p_{xb} & 2p_{xb} \\ 2p_{ya} & 2p_{yb} \end{bmatrix}_R$ |
| $C_i(min)$ | 0.240 94 | 0.181 73 | 0.138 44 | 0.125 86 |

Comparing Tables 12.5 and 12.6 with those for $N_2$, Tables 12.2 and 12.3, we see that CO prefers hybrid orbitals to a somewhat greater extent. The differences are not great, however. The EGSO weights shown in Table 12.7 display a behavior rather different from those for $N_2$. In this case the principal configuration is the same for both sorts of measure. The smaller EGSO weights are different, however.

### 12.1.3 BF

The pattern of tables for BF follows the earlier treatments in the chapter. Table 12.8 shows the four most important structures in the wave function of BF as determined by the magnitude of the coefficients for standard tableaux functions and HLSP functions. Table 12.9 shows the same information for the $\sigma$ AOs formed into $s$–$p$ hybrids. We use the "$h_{ox}$" or "$h_{ix}$" symbols as before. In this case, all of the principal structures except number 4 for the hybrid AOs have no more than one unpaired set of orbitals. Therefore, the coefficients for the standard tableaux functions and the HLSP functions differ in that case only. The hybrid orbital arrangement is again preferred, but the difference is only somewhat greater than that for CO. The principal configuration is definitely a single $\sigma$ bond between the two atoms. When interpreted as a configuration of BF, the most important one from CO changes to the ionic sort.

Table 12.7. *CO: EGSO weights (standard tableaux functions) for spherical AOs, upper group, and s–p hybrids, lower group. These are weights for whole symmetry functions rather than individual tableaux. It should be recalled from Chapter 6 that the detailed forms of symmetry functions are dependent on the particular arrangement of the orbitals in the tableaux and are frequently nonintuitive.*

|  | 1 | 2 | 3 | 4 |
|---|---|---|---|---|
| Num. | 2 | 1 | 2 | 2 |
| STF | $\begin{bmatrix} 2s_a & 2s_a \\ 2s_b & 2s_b \\ 2p_{yb} & 2p_{yb} \\ 2p_{za} & 2p_{zb} \\ 2p_{xa} & 2p_{xb} \end{bmatrix}$ | $\begin{bmatrix} 2s_a & 2s_a \\ 2s_b & 2s_b \\ 2p_{zb} & 2p_{zb} \\ 2p_{xa} & 2p_{xb} \\ 2p_{ya} & 2p_{yb} \end{bmatrix}$ | $\begin{bmatrix} 2s_b & 2s_b \\ 2p_{zb} & 2p_{zb} \\ 2p_{yb} & 2p_{yb} \\ 2s_a & 2p_{za} \\ 2p_{xa} & 2p_{xb} \end{bmatrix}$ | $\begin{bmatrix} 2s_a & 2s_a \\ 2s_b & 2s_b \\ 2p_{zb} & 2p_{zb} \\ 2p_{xa} & 2p_{xa} \\ 2p_{yb} & 2p_{yb} \end{bmatrix}$ |
| Wt | 0.550 208 | 0.140 467 | 0.105 266 | 0.033 16 |
| Num. | 2 | 1 | 2 | 1 |
| STF | $\begin{bmatrix} h_{oa} & h_{oa} \\ h_{ob} & h_{ob} \\ 2p_{yb} & 2p_{yb} \\ h_{ia} & h_{ib} \\ 2p_{xa} & 2p_{xb} \end{bmatrix}$ | $\begin{bmatrix} h_{oa} & h_{oa} \\ h_{ob} & h_{ob} \\ h_{ib} & h_{ib} \\ 2p_{xa} & 2p_{xb} \\ 2p_{ya} & 2p_{yb} \end{bmatrix}$ | $\begin{bmatrix} h_{ob} & h_{ob} \\ h_{ib} & h_{ib} \\ 2p_{yb} & 2p_{yb} \\ h_{oa} & h_{ia} \\ 2p_{xa} & 2p_{xb} \end{bmatrix}$ | $\begin{bmatrix} h_{oa} & h_{oa} \\ h_{ob} & h_{ob} \\ 2p_{xb} & 2p_{xb} \\ 2p_{yb} & 2p_{yb} \\ h_{ia} & h_{ib} \end{bmatrix}$ |
| Wt | 0.637 66 | 0.099 84 | 0.041 74 | 0.031 57 |

Table 12.8. *BF: The most important terms in the wave function when spherical AOs are used as determined by the magnitudes of the coefficients. Results for standard tableaux and HLSP functions are the same.*

|  | 1 | 2 | 3 | 4 |
|---|---|---|---|---|
| Num. | 1 | 2 | 1 | 1 |
| STF or HLSP | $\begin{bmatrix} 2s_a & 2s_a \\ 2s_b & 2s_b \\ 2p_{xb} & 2p_{xb} \\ 2p_{yb} & 2p_{yb} \\ 2p_{za} & 2p_{zb} \end{bmatrix}$ | $\begin{bmatrix} 2s_a & 2s_a \\ 2s_b & 2s_b \\ 2p_{zb} & 2p_{zb} \\ 2p_{yb} & 2p_{yb} \\ 2p_{xa} & 2p_{xb} \end{bmatrix}$ | $\begin{bmatrix} 2s_a & 2s_a \\ 2s_b & 2s_b \\ 2p_{zb} & 2p_{zb} \\ 2p_{xb} & 2p_{xb} \\ 2p_{yb} & 2p_{yb} \end{bmatrix}$ | $\begin{bmatrix} 2s_b & 2s_b \\ 2p_{zb} & 2p_{zb} \\ 2p_{xb} & 2p_{xb} \\ 2p_{yb} & 2p_{yb} \\ 2s_a & 2p_{za} \end{bmatrix}$ |
| $C_i(min)$ | 0.277 78 | $-0.224\,278$ | $-0.217\,808$ | $-0.135\,735$ |

For spherical AOs it is not among the first four, but appears in the eighth position with coefficient of 0.114 16, and for hybrids it is the eighth one down with a coefficient of 0.120 80. We therefore predict that, quantitatively, there is less $\pi$ back-bonding in BF than in CO. For neither arrangement of orbitals is the triply bonded structure of $N_2$ important for BF.

Table 12.9. *BF: The most important terms in the wave function when hybrid AOs are used as determined by the magnitudes of the coefficients. Results for standard tableaux and HLSP functions are the same.*

| | 1 | 2 | 3 | 4 |
|---|---|---|---|---|
| Num. | 1 | 1 | 2 | 2 |
| STF or HLSP | $\begin{bmatrix} h_{oa} & h_{oa} \\ h_{ob} & h_{ob} \\ 2p_{xb} & 2p_{xb} \\ 2p_{yb} & 2p_{yb} \\ h_{ia} & h_{ib} \end{bmatrix}$ | $\begin{bmatrix} h_{oa} & h_{oa} \\ h_{ob} & h_{ob} \\ h_{ib} & h_{ib} \\ 2p_{xb} & 2p_{xb} \\ 2p_{yb} & 2p_{yb} \end{bmatrix}$ | $\begin{bmatrix} h_{oa} & h_{oa} \\ h_{ob} & h_{ob} \\ h_{ib} & h_{ib} \\ 2p_{yb} & 2p_{yb} \\ 2p_{xa} & 2p_{xb} \end{bmatrix}$ | $\begin{bmatrix} h_{ob} & h_{ob} \\ h_{ib} & h_{ib} \\ 2p_{yb} & 2p_{yb} \\ h_{oa} & h_{ia} \\ 2p_{xa} & 2p_{xb} \end{bmatrix}$ |
| $C_i^{stf}(min)$ | 0.335 49 | 0.260 92 | 0.228 79 | 0.160 75 |
| $C_i^{hlsp}(min)$ | 0.335 49 | 0.260 92 | 0.228 79 | 0.147 62 |

Table 12.10. *BF: EGSO weights (standard tableaux functions) for spherical AOs, upper group, and hybrid AOs, lower group. These are weights for whole symmetry functions.*

| | 1 | 2 | 3 | 4 |
|---|---|---|---|---|
| Num. | 2 | 1 | 1 | 2 |
| STF | $\begin{bmatrix} 2s_a & 2s_a \\ 2s_b & 2s_b \\ 2p_{zb} & 2p_{zb} \\ 2p_{yb} & 2p_{yb} \\ 2p_{xa} & 2p_{xb} \end{bmatrix}$ | $\begin{bmatrix} 2s_a & 2s_a \\ 2s_b & 2s_b \\ 2p_{xb} & 2p_{xb} \\ 2p_{yb} & 2p_{yb} \\ 2p_{za} & 2p_{zb} \end{bmatrix}$ | $\begin{bmatrix} 2s_b & 2s_b \\ 2p_{zb} & 2p_{zb} \\ 2p_{xb} & 2p_{xb} \\ 2p_{yb} & 2p_{yb} \\ 2s_a & 2p_{za} \end{bmatrix}$ | $\begin{bmatrix} 2s_b & 2s_b \\ 2p_{zb} & 2p_{zb} \\ 2p_{yb} & 2p_{yb} \\ 2s_a & 2p_{za} \\ 2p_{xa} & 2p_{xb} \end{bmatrix}$ |
| Wt | 0.464 83 | 0.302 50 | 0.082 13 | 0.028 11 |
| Num. | 1 | 2 | 2 | 1 |
| STF | $\begin{bmatrix} h_{oa} & h_{oa} \\ h_{ob} & h_{ob} \\ 2p_{xb} & 2p_{xb} \\ 2p_{yb} & 2p_{yb} \\ h_{ia} & h_{ib} \end{bmatrix}$ | $\begin{bmatrix} h_{oa} & h_{oa} \\ h_{ob} & h_{ob} \\ h_{ib} & h_{ib} \\ 2p_{yb} & 2p_{yb} \\ 2p_{xa} & 2p_{xb} \end{bmatrix}$ | $\begin{bmatrix} h_{ob} & h_{ob} \\ h_{ib} & h_{ib} \\ 2p_{yb} & 2p_{yb} \\ h_{oa} & h_{ia} \\ 2p_{xa} & 2p_{xb} \end{bmatrix}$ | $\begin{bmatrix} h_{ob} & h_{ob} \\ h_{ia} & h_{ia} \\ 2p_{xb} & 2p_{xb} \\ 2p_{yb} & 2p_{yb} \\ h_{oa} & h_{ib} \end{bmatrix}$ |
| Wt | 0.486 57 | 0.252 67 | 0.078 54 | 0.036 12 |

The EGSO weights shown in Table 12.10 for the two orbital arrangements display an interesting switch. For the spherical AOs a cylindrically averaged $\pi$ bond is the principal configuration and the $\sigma$ bond is the second one. The hybrid AOs show the opposite order with the $\sigma$ bond structure relatively more strongly favored than in the spherical case. In both cases there is considerable competition between the two bond types, and the VB prediction is that they are strongly mixed in the molecule. We defer a discussion of the dipole moment until later.

Table 12.11. *BeNe: The most important terms in the wave function when spherical AOs are used, as determined by the magnitudes of the coefficients. Results for standard tableaux and HLSP functions are the same for these terms in the wave function.*

| | 1 | 2 | 3 | 4 |
|---|---|---|---|---|
| Num. | 1 | 2 | 1 | 1 |
| STF or HLSP | $\begin{bmatrix} 2s_a & 2s_a \\ 2s_b & 2s_b \\ 2p_{zb} & 2p_{zb} \\ 2p_{xb} & 2p_{xb} \\ 2p_{yb} & 2p_{yb} \end{bmatrix}$ | $\begin{bmatrix} 2s_b & 2s_b \\ 2p_{zb} & 2p_{zb} \\ 2p_{xa} & 2p_{xa} \\ 2p_{xb} & 2p_{xb} \\ 2p_{yb} & 2p_{yb} \end{bmatrix}$ | $\begin{bmatrix} 2s_a & 2s_a \\ 2s_b & 2s_b \\ 2p_{xb} & 2p_{xb} \\ 2p_{yb} & 2p_{yb} \\ 2p_{za} & 2p_{zb} \end{bmatrix}$ | $\begin{bmatrix} 2s_b & 2s_b \\ 2p_{za} & 2p_{za} \\ 2p_{zb} & 2p_{zb} \\ 2p_{xb} & 2p_{xb} \\ 2p_{yb} & 2p_{yb} \end{bmatrix}$ |
| $C_i(min)$ | 0.879 46 | −0.188 65 | −0.186 14 | −0.144 05 |

Table 12.12. *BeNe: The most important terms in the wave function when s–p hybrid AOs are used as determined by the magnitudes of the coefficients. Results for standard tableaux and HLSP functions are the same for these terms in the wave function.*

| | 1 | 2 | 3 | 4 |
|---|---|---|---|---|
| Num. | 1 | 1 | 1 | 2 |
| STF or HLSP | $\begin{bmatrix} h_{ob} & h_{ob} \\ h_{ib} & h_{ib} \\ 2p_{xb} & 2p_{xb} \\ 2p_{yb} & 2p_{yb} \\ h_{oa} & h_{ia} \end{bmatrix}$ | $\begin{bmatrix} h_{oa} & h_{oa} \\ h_{ob} & h_{ob} \\ h_{ib} & h_{ib} \\ 2p_{xb} & 2p_{xb} \\ 2p_{yb} & 2p_{yb} \end{bmatrix}$ | $\begin{bmatrix} h_{ob} & h_{ob} \\ h_{ia} & h_{ia} \\ h_{ib} & h_{ib} \\ 2p_{xb} & 2p_{xb} \\ 2p_{yb} & 2p_{yb} \end{bmatrix}$ | $\begin{bmatrix} h_{ob} & h_{ob} \\ h_{ib} & h_{ib} \\ 2p_{xb} & 2p_{xb} \\ 2p_{ya} & 2p_{ya} \\ 2p_{yb} & 2p_{yb} \end{bmatrix}$ |
| $C_i(min)$ | 0.718 54 | 0.429 47 | 0.298 39 | −0.188 64 |

### 12.1.4 BeNe

When we arrive at BeNe in our series we expect no real electron pair bond between the two atoms, but we provide the same sorts of tables as before. Table 12.11 shows the four most important structures in the wave function of BF as determined by the magnitude of the coefficients for standard tableaux functions or HLSP functions. Table 12.12 shows the same information for the $\sigma$ AOs formed into s–p hybrids. The symbols "$h_{ox}$" or "$h_{ix}$" are used as before. In this case, where none of the principal structures has more than one pair of unpaired orbitals, there is no difference in the coefficients between the standard tableaux functions and the HLSP functions. For BeNe the spherical AO arrangement is definitely preferred. Examination of Table 12.12 shows an unusual inner–outer hybrid pairing on Be in the principal configuration. This pairing is not a good substitute for the $2s^2$ ground

Table 12.13. *BeNe: EGSO weights (standard tableaux functions) for spherical AOs. These are weights for whole symmetry functions.*

|  | 1 | 2 | 3 | 4 |
|---|---|---|---|---|
| Num. | 1 | 2 | 1 | 1 |
| STF | $\begin{bmatrix} 2s_a & 2s_a \\ 2s_b & 2s_b \\ 2p_{zb} & 2p_{zb} \\ 2p_{xb} & 2p_{xb} \\ 2p_{yb} & 2p_{yb} \end{bmatrix}$ | $\begin{bmatrix} 2s_b & 2s_b \\ 2p_{zb} & 2p_{zb} \\ 2p_{xa} & 2p_{xa} \\ 2p_{xb} & 2p_{xb} \\ 2p_{yb} & 2p_{yb} \end{bmatrix}$ | $\begin{bmatrix} 2s_a & 2s_a \\ 2s_b & 2s_b \\ 2p_{xb} & 2p_{xb} \\ 2p_{yb} & 2p_{yb} \\ 2p_{za} & 2p_{zb} \end{bmatrix}$ | $\begin{bmatrix} 2s_b & 2s_b \\ 2p_{za} & 2p_{za} \\ 2p_{zb} & 2p_{zb} \\ 2p_{xb} & 2p_{xb} \\ 2p_{yb} & 2p_{yb} \end{bmatrix}$ |
| Wt | 0.845 66 | 0.078 54 | 0.035 88 | 0.020 73 |
| Num. | 1 | 1 | 1 | 2 |
| STF | $\begin{bmatrix} h_{ob} & h_{ob} \\ h_{ib} & h_{ib} \\ 2p_{xb} & 2p_{xb} \\ 2p_{yb} & 2p_{yb} \\ h_{oa} & h_{ia} \end{bmatrix}$ | $\begin{bmatrix} h_{oa} & h_{oa} \\ h_{ob} & h_{ob} \\ h_{ib} & h_{ib} \\ 2p_{xb} & 2p_{xb} \\ 2p_{yb} & 2p_{yb} \end{bmatrix}$ | $\begin{bmatrix} h_{ob} & h_{ob} \\ h_{ia} & h_{ia} \\ h_{ib} & h_{ib} \\ 2p_{xb} & 2p_{xb} \\ 2p_{yb} & 2p_{yb} \end{bmatrix}$ | $\begin{bmatrix} h_{ob} & h_{ob} \\ h_{ib} & h_{ib} \\ 2p_{xa} & 2p_{xa} \\ 2p_{xb} & 2p_{xb} \\ 2p_{yb} & 2p_{yb} \end{bmatrix}$ |
| Wt | 0.573 15 | 0.208 39 | 0.093 72 | 0.078 54 |

state of Be and leads to the somewhat smaller value of the coefficient compared to the primary structure in the case of spherical AOs. Because of the column antisymmetry of the standard tableaux functions the hybrids on Ne do not cause a similar difficulty.

Comparing the two sets of weights in Table 12.13 shows the same phenomenon. The principal spherical AO structure represents over 80% of the total wave function while the weights for the hybrid structures fall off more slowly. It addition it will be observed that the principal terms in the wave functions and the EGSO weights are completely parallel in the case of BeNe. This is in contrast to other members in this series of molecules.

We see that the third function in any of the spherical AO series is an ionic structure equivalent to the principal configuration for BF and thus represents one $\sigma$ bond. This is a relatively minor constituent of the wave function, but, nevertheless, has a surprisingly large coefficient. It is possible that this sort of term is overemphasized in the STO3G basis, since it predicts an improbably short bond between Be and Ne, judged by the value obtained with the higher-quality 6-31G* basis.[1] As a test of this conjecture, a recalculation of the STO3G structures at the 6-31G* equilibrium distance reduces the importance of this ionic structure to the fourth place with an EGSO weight of 0.4%. The $\pi$ back-bonding

---

[1] This is very likely a manifestation of *basis set superposition error* that occurs frequently in MO calculations, also.

Table 12.14. *Statistics on 6-31G\* calculations for N$_2$, CO, BF, and BeNe. The dipole moments are also given. See Section 12.3 for a discussion of the signs of the moments.*

| Molecule | $N_I{}^a$ | $N_S{}^b$ | Moment (D) | |
|----------|-----------|-----------|------------|------|
| | | | Calc. | Exp. |
| N$_2$ | 6964 | 1812 | 0.0 | 0.0 |
| CO | 5736 | 2986 | −0.087 | −0.122 |
| BF | 3166 | 1680 | −1.084 | −0.5$^c$ |
| BeNe | 1210 | 672 | −0.312 | ? |

$^a$ The number of basis functions involved in $^1\Sigma_g^+$ or $^1\Sigma^+$ symmetry functions.
$^b$ The number of symmetry functions supported.
$^c$ For $v = 0$.

structure at $\approx 0.01\%$ is even less important in this molecule. It appears that VB theory predicts there to be no electron pair bonds between the two atoms here. The minimum in the internuclear potential curve is due to Van der Waals interactions. In spite of this the molecule has a small dipole moment, which we discuss below.

## 12.2 Quantitative results from a 6-31G\* basis

In Chapter 11 we described calculations using the occupied AOs in a full MCVB with added configurations involving single excitations into all of the atomic virtual orbitals excepting the $d_\delta$. We give the values for $D_e$ and $R_m$ in Table 12.1. The results for N$_2$ are the same, of course, as those in Table 11.1. As was the case with the homonuclear molecules in Chapter 11, we again see that the calculated energy for N$_2$ is the farthest from experiment for the known values. There seems at the moment no good explanation for this. Nevertheless, the higher-quality basis gives closer agreement with experiment. In Table 12.14 we present statistics for the number of $n$-electron basis functions involved in the calculations.

Apparently there are no experimental data on BeNe. If we fit a Morse function to the parameters we obtain for the dissociation curve, it is estimated that there would be 14–15 bound vibrational states for this Van der Waals molecule. Thus, VB theory predicts the existence of stable gaseous BeNe, if it is cold enough, since $D_e$ is only $2kT$ for room temperature.

As stated above, we consider the dipole moments of the heteronuclear molecules in the next section, but we give in Table 12.14 the dipole moments at the equilibrium geometry and determined with the 6-31G\* basis.

## 12.3 Dipole moments of CO, BF, and BeNe

Elementary discussions define the electronegativity of an atom as a measure of its ability to attract electrons to itself. Several authors, Pauling[50], Mulliken[51], and Allen[52], have devised quantitative values as a measure of this ability. Such elementary discussions usually emphasize the connection between the dipole moments of heteronuclear bonds and the comparative electronegativities of the atoms involved. In particular the expectation is that the electronegativity difference should tell the direction of the moment. In general, this idea works well with many diatomic molecules that have single bonds between the atoms. Examples are hydrogen halides and (gaseous) alkali halides. Discussions of LiH and LiF representing this sort of system are in Chapter 8. There are, nevertheless, a number of diatomic molecules that have an anomalous direction of the dipole moment between different atoms. CO is probably the most notorious of these anomalies but others are known. Huzinaga *et al.*[53] have examined a number of these and describe the effects in terms of MO theories. The interested reader is referred to the article for the details, since this work stresses VB analyses of chemical phenomena.

### *12.3.1 Results for 6-31G\* basis*

Figure 12.1 shows the dipole moment functions in terms of internuclear distance of CO, BF, and BeNe, calculated with our conventional 6-31G\* basis arrangement.

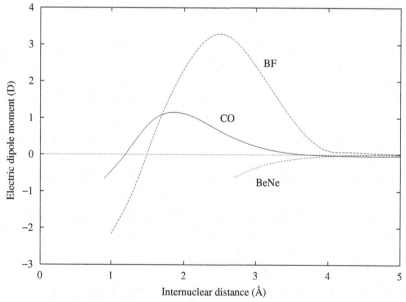

Figure 12.1. The dipole moment functions for CO, BF, BeNe calculated at a number of distances with the conventional 6-31G\* basis arrangement.

The curves for CO and BF show the form typical of these systems as is emphasized in Ref. [53]. In our discussion of LiF in Section 8.3 we emphasized how the nature of a wave function could change from ionic to covalent with a change in internuclear distance. Here again we appear to have the "signature" of this sort of phenomenon: the change of sign of the moment at internuclear distances around 1.0–1.5 Å strongly suggests the interplay of two effects where the winning one changes fairly rapidly with distance. On the other hand the sign of BeNe does not change and this suggests that one of these effects is absent in this molecule.

From the signs on the moments and our work in Chapters 2 and 8 we interpret these curves as follows (for the interpretation of the signs the reader is reminded that all three of our systems are oriented with the *less* electronegative atom in the positive $z$-direction).

1. At internuclear distances intermediate, but greater than equilibrium, the familiar ideas of electronegativity win out, and the more electronegative atom has an excess of negative charge. At the maxima the charge on O in CO is around $-0.29|e|$ and on F in BF $-0.70|e|$. It is not surprising that in BF the effect is larger. No legitimate argument would suggest that Ne has any sort of negative ion propensity, and we do not see a maximum in that curve.

2. When systems are pushed together, nonbonded electrons, on the other hand, tend to retreat toward the system that has the more diffuse orbitals. In this case that is C, B, or Be. Since the nonbonded electrons are generally in orbitals less far out, this effect occurs at closer distances and, according to our calculations, wins out at equilibrium distances for CO and BF. This is the only effect for BeNe, and the moment is in the same direction at all of the distances we show. This retreat of electrons is definitely a result of the Pauli exclusion principle.

Both sorts of physical effects tend to fall off exponentially as the distance between the atoms increases – the dipole moment must go to zero asymptotically. A close examination of the CO results shows that the moment goes to very small negative values again around 4.0 Å. Whether this is real is difficult to decide without further calculations. It might be that the Pauli exclusion effect wins again at these distances, the result might be different for a still larger basis. Also, Gaussian basis functions can cause troubles at larger distances because individually they really fall off much too rapidly with distance.

### 12.3.2 Difficulties with the STO3G basis

We also calculated the dipole moment functions for CO, BF, and BeNe with an STO3G basis, and it can be seen in Fig. 12.2 that there are real difficulties with the minimal basis. We have argued that the numerical value and sign of the electric

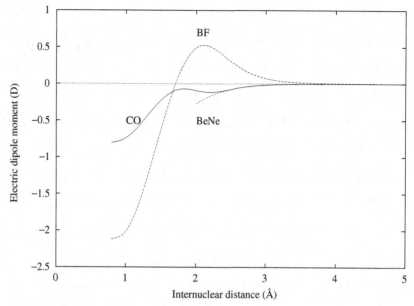

Figure 12.2. The dipole moment functions for CO, BF, BeNe calculated at a number of distances with the full valence minimal STO3G basis.

dipole moment in these molecules is the result of a balance between two opposing effects. When we pass to the STO3G basis, this balance is thrown off.[2] It is instructive, nevertheless, to see that STO3Gs reproduce the Pauli exclusion effect better than the formation of a partial negative ion at the more electronegative end of the molecule. This is expected. The more diffuse 6-31G* basis, with its capability for allowing the AOs to breathe and polarize can much better represent the negative ions.

---

[2] This is in spite of the qualitatively reasonable energies the basis yields. Such an outcome is a familiar one, however – the energy is the result of a variational calculation and is expected to be produced to higher order than quantities like the electric moment. In addition, the minimal basis does better in the region of the minimum and asymptotically than elsewhere. Thus, $D_e$ may not suffer too greatly.

# 13

# Methane, ethane and hybridization

In Chapter 11 we discussed the properties of the atoms in the second row of the periodic table and how these might influence molecules formed from them. We focus on carbon in this chapter and examine how the bonding changes through the series CH, $CH_2$, $CH_3$, and $CH_4$. The first three of these are known only spectroscopically, in matrix isolation, or as reaction intermediates, but many of their properties have been determined. The reader will recall that carbon exhibits relatively low-energy excited valence configurations. For carbon the excitation energy is around 4 eV, and among the atoms discussed in Chapter 11, only boron has a lower excitation energy. If this excited configuration is to have an important role in the bonding, the energy to produce the excitation must be paid back by the energy of formation of the bond or bonds. We shall see that VB theory predicts this happens between CH and $CH_2$. After our discussion of these single carbon compounds, we will consider ethane, $CH_3CH_3$, as an example for dealing with larger hydrocarbons.

## 13.1 CH, $CH_2$, $CH_3$, and $CH_4$

### 13.1.1 STO3G basis

We first give calculations of these four molecules with an STO3G basis. The total energies and first bond dissociation energies are collected in Table 13.1. We see that, even with the minimal basis, the bond energies are within 0.4 eV of the experimental values except for $CH_3$, which has considerable uncertainty. The calculated values tend to be smaller, as expected for a minimal VB treatment.

We now give a discussion of each of the molecules, first considering the atomic structure of the carbon atom and attempting to predict the bonding pattern. The predictions are followed by the results of the STO3G calculations.

### Table 13.1. *CH$_n$ STO3G energies.*

| $n$ | | Energy (au) | Dissociation | Energy (eV) | Exp. (eV)[a] |
|-----|---|-------------|--------------|-------------|-----------|
| 0 | C ($^3P$) | −37.438 66 | | | |
| 1 | CH ($^2\Pi$) | −38.050 28 | $D_{C-H}$ | 3.065 | 3.465 |
| 2 | CH$_2$ ($^3B_1$) | −38.693 63 | $D_{CH-H}$ | 3.901 | 4.33 |
| 3 | CH$_3$ ($^2A_2''$) | −39.338 38 | $D_{CH_2-H}$ | 3.939 | ≤4.90 |
| 4 | CH$_4$ ($^1A_1$) | −39.989 73 | $D_{CH_3-H}$ | 4.118 | 4.406 |

[a] See Refs. [49, 54].

### Table 13.2. *Principal standard tableaux functions for CH at the equilibrium internuclear distance. This is the x-component of a π-pair.*

| | 1 | 2 | 3 | 4 |
|---|---|---|---|---|
| Num.[a] | 1 | 1 | 1 | 1 |
| Tab.[b] | $\begin{bmatrix} 2s & 2s \\ 1s & 2p_z \\ 2p_x & \end{bmatrix}$ | $\begin{bmatrix} 2s & 2s \\ 1s & 2p_x \\ 2p_z & \end{bmatrix}$ | $\begin{bmatrix} 2s & 2s \\ 2p_z & 2p_z \\ 2p_x & \end{bmatrix}$ | $\begin{bmatrix} 2p_z & 2p_z \\ 2s & 1s \\ 2p_x & \end{bmatrix}$ |
| $C_i(min)$ | 0.729 684 53 | −0.320 400 21 | 0.227 852 03 | 0.168 832 87 |

[a] The number of terms in the symmetry function that is generated from the tableau shown. (See text.)
[b] These tableau symbols exclude the core orbitals.

### CH

The $^3P$ (we call it $^3P(1)$) ground state of the C atom has two unpaired $p$ electrons. When an H atom approaches, it should be able to form an electron pair bond with one of these orbitals, while the other would remain unpaired. This scenario leads to the expectation that CH should have a $^2\Pi$ ground state. We have commented on the possible involvement of the excited C $^5S$ state, but symmetry prohibits such mixing here. There is a higher energy $^3P(2)$ valence state that is allowed to interact through symmetry.

There are 75 standard tableaux functions in a full valence treatment, but only 36 are $\Pi$ states, half being $x$-components and half $y$-components. The variation problem therefore has two $18 \times 18$ matrices. The principal standard tableaux functions in the wave function are shown in Table 13.2. The predominant term in the wave function clearly involves the C atom in its $^3P(1)$ state. The calculated dipole

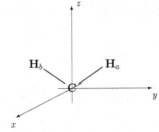

Figure 13.1. Orientation of the CH₂ diradical.

moment at the equilibrium separation for this basis is 1.4983 D with H the positive end.

## CH₂

The methylene radical has enjoyed a certain notoriety concerning the nature of the ground state. It is now known to be in a triplet state with a bent geometry. This is perhaps not what is expected if we just think of an H atom interacting with the remaining unpaired $p$ orbital of CH, an outcome that should lead to a singlet state of some geometry. At this stage in our series we will see, however, that the excited $^5S$ state becomes dominant in the wave function. A quintet state coupled with two doublet H atoms can have no lower multiplicity than triplet. In Fig. 13.1 we show the orientation of the CH₂ diradical in a Cartesian coordinate system and assume $C_{2v}$ symmetry.

With six electrons and six orbitals in a full valence calculation we expect 189 standard tableaux functions. These support 51 $^3B_1$ symmetry functions that, however, involve a total of only 97 standard tableaux functions out of the possible 189. Table 13.3 shows the principal terms in the wave function for the equilibrium geometry.

The principal standard tableaux function is

$$T_1^{AO} = \begin{bmatrix} 2s & 1s_a \\ 2p_x & 1s_b \\ 2p_y & \\ 2p_z & \end{bmatrix}, \tag{13.1}$$

where the subscripts on the 1s orbitals are associated with the corresponding subscripts on the H atoms in Fig. 13.1, and the $1s_a$ orbital is on the positive $y$ side of the $x$–$z$ plane with $1s_b$ on the other side. We add a superscript "$AO$" to the tableau

Table 13.3. *Principal standard tableaux functions for CH$_2$ at the equilibrium internuclear geometry.*

| | 1 | 2 | 3 | 4 |
|---|---|---|---|---|
| Num.[a] | 1 | 2 | 2 | 1 |
| Tab.[b] | $\begin{bmatrix} 2s & 1s_a \\ 2p_x & 1s_b \\ 2p_y \\ 2p_z \end{bmatrix}$ | $\begin{bmatrix} 2p_y & 2p_y \\ 2s & 1s_a \\ 2p_x \\ 2p_z \end{bmatrix}$ | $\begin{bmatrix} 2s & 2s \\ 2p_x & 1s_b \\ 2p_z \\ 1s_a \end{bmatrix}$ | $\begin{bmatrix} 2s & 2s \\ 2p_x & 2p_y \\ 1s_a \\ 1s_b \end{bmatrix}$ |
| $C_i(min)$ | 0.379 324 12 | 0.092 437 13 | −0.091 619 94 | −0.088 823 52 |

[a] The number of terms in the symmetry function that is generated from the tableau shown. (See text).
[b] These tableau symbols exclude the core orbitals.

symbol to distinguish it from tableaux we introduce later that have hybrid orbitals in them.

As we have pointed out many times previously, the columns of the standard tableaux functions are antisymmetrized, and the orbitals in a column may be replaced by any linear combination of them with no more than a change of an unimportant overall constant. In this case, consider a linear combination that has two hybrid orbitals that point directly at the H atoms in accord with Pauling's principle of maximum overlap. Using the parameter $\phi$ we have three orthonormal hybrids

$$\left.\begin{aligned} h_a &= \cos(\theta)(2s) + \sin(\theta)[\sin(\phi/2)(2p_y) + \cos(\phi/2)(2p_z)], \\ h_b &= \cos(\theta)(2s) + \sin(\theta)[-\sin(\phi/2)(2p_y) + \cos(\phi/2)(2p_z)], \\ h_z &= \frac{\cos(\phi/2)\sin(\theta)(2s) - \cos(\theta)(2p_z)}{\sqrt{\cos^2(\theta) + \cos^2(\phi/2)\sin^2(\theta)}}, \end{aligned}\right\} \qquad (13.2)$$

where $\phi\,(>\pi/2)^1$ is the angle between these two hybrids, and

$$\theta = \mathrm{arccot}\left[\sqrt{-\cos(\phi)}\right].$$

We keep the $2p_x$ orbital unchanged. This set is a variant of the canonical $sp^2$ hybrid set in which, however, all three orbitals are not symmetrically equivalent. If $\phi$ is the H–C–H angle, $h_a$ and $h_b$ point directly at the H atoms. Because of the invariance

---

[1] Hybrids consisting of $s$ and $p$ orbitals without this angle restriction can be complex. We are not interested in such cases.

Table 13.4. *Energies for $T_1^{AO}$ and $T_1^{hR}$ as a function of hybrid angle.*

| Hybrid angle | $\dfrac{\langle T_1^{AO}\|H\|T_1^{AO}\rangle}{\langle T_1^{AO}\|T_1^{AO}\rangle}$ au | $\dfrac{\langle T_1^{hR}\|H\|T_1^{hR}\rangle}{\langle T_1^{hR}\|T_1^{hR}\rangle}$ au |
|---|---|---|
| 120.0 | 38.479 436 | −38.595 532 |
| 130.0 | −38.479 436 | −38.599 825 |
| 140.0 | −38.479 436 | −38.597 295 |
| 150.0 | −38.479 436 | −38.590 391 |

of the function in Eq. (13.1), we obtain

$$T_1^{AO} = \begin{bmatrix} h_a & 1s_a \\ h_b & 1s_b \\ h_z & \\ 2p_x & \end{bmatrix}. \tag{13.3}$$

We have written out these hybrids, but the reader should realize that the eight-electron wave function (including the $1s^2$) based upon the standard tableaux function of Eq. (13.3) has an energy expectation value independent[2] of the angle parameter used in the hybrids, so long as $\phi > \pi/2$. There are, however, nine standard tableaux functions for the orbital configuration in Eq. (13.3). These may be combined into five other combinations of $^3B_1$ symmetry. The $T_1^{AO}$ above is the only one that shows the invariance to hybrid angle. When we combine all five in a wave function, the energy does depend upon the hybrid orbital directions. Nevertheless, as we add more and more structures to the wave function, we eventually arrive at a full calculation, and the energy is again invariant to the hybrid orbital directions. Thus the principal of maximum overlap has a meaning only for wave functions that do not involve a linear combination of all possible structures for the underlying AO basis.

Some further numerical examples are illuminating when we compare the standard tableaux function results with those of HLSP functions. We define

$$T_1^{hR} = \begin{bmatrix} h_a & 1s_a \\ h_b & 1s_b \\ 2p_x & \\ h_z & \end{bmatrix}_R. \tag{13.4}$$

Consider the energies in Table 13.4, where we see that the energy of $T_1^{hR}$ varies up and down around 100–200 meV in the angle range shown, while that of $T_1^{AO}$ is

---

[2] This does not mean that the energy of this $T_1^h$ is independent of the actual angle in the molecule. Among other things, the nuclear repulsion energy depends upon the distance between the H atoms.

constant. Using the methods of Chapter 5, we may write $T_1^{hR}$ in terms of standard tableaux functions:

$$
T_1^{hR} = \frac{12}{5} \begin{bmatrix} h_a & 1s_a \\ h_b & 1s_b \\ h_z & \\ 2p_x & \end{bmatrix} - \frac{6}{5} \begin{bmatrix} h_a & 1s_a \\ h_b & h_z \\ 1s_b & \\ 2p_x & \end{bmatrix} + \frac{6}{5} \begin{bmatrix} h_a & 1s_a \\ h_b & 2p_x \\ 1s_b & \\ h_z & \end{bmatrix}
$$

$$
- \frac{6}{5} \begin{bmatrix} h_a & h_b \\ 1s_a & 1s_b \\ h_z & \\ 2p_x & \end{bmatrix} + \frac{3}{5} \begin{bmatrix} h_a & h_b \\ 1s_a & h_z \\ 1s_b & \\ 2p_x & \end{bmatrix} - \frac{3}{5} \begin{bmatrix} h_a & h_b \\ 1s_a & 2p_x \\ 1s_b & \\ h_z & \end{bmatrix}
$$

$$
- \frac{9}{5} \begin{bmatrix} h_a & 1s_b \\ 1s_a & h_z \\ h_b & \\ 2p_x & \end{bmatrix} + \frac{9}{5} \begin{bmatrix} h_a & 1s_b \\ 1s_a & 2p_x \\ h_b & \\ h_z & \end{bmatrix} - \frac{6}{5} \begin{bmatrix} h_a & h_z \\ 1s_a & 2p_x \\ h_b & \\ 1s_b & \end{bmatrix} . \quad (13.5)
$$

We see immediately that $T_1^{AO}$, the standard tableaux function invariant to the hybrid angles, is actually the largest term in $T_1^{hR}$, but not overwhelmingly so. The others all depend on the hybrid angles and, therefore, so does $T_1^{hR}$. We may also note that using hybrid orbitals $T_1^{hR}$ has a lower energy by $\approx 3$ eV, but, as seen from Table 13.1, the full calculation, with either sort of basis, is still more stable by another $\approx 3$ eV.

The calculated value of the dipole moment is 0.6575 D for this basis with the charge positive at the H-atom end of the bonds.

## $CH_3$

Adding an H atom to $CH_2$ might be expected to do little more than regularize the hybrids we gave in Eq. (13.2), converting them to a canonical $sp^2$ set. With this we expect a planar doublet system. Whether the molecule is really planar is difficult to judge from qualitative considerations. Calculations and experiment bear out the planarity, however.

A full valence orbital VB calculation in this basis involves 784 standard tableaux functions, of which only 364 are involved in 68 $^2A_2''$ symmetry functions. For $CH_3$ we present the results in terms of $sp^2$ hybrids. This has no effect on the energy, of course. We show the principal standard tableaux functions in Table 13.5. The molecule is oriented with the $C_3$-axis along the $z$-axis and one of the H atoms on the $x$-axis. The three trigonal hybrids are oriented towards the H atoms. The "$x$" subscript on the orbital symbols in Table 13.5 indicates the functions on the $x$-axis, the "$a$" subscript those 120° from the first set, and the "$b$" subscript those 240° from the first set.

Table 13.5. *Principal standard tableaux function structures for CH$_3$ at equilibrium bond distances.*

| | 1 | 2 | 3 | 4 |
|---|---|---|---|---|
| Num.[a] | 1 | 1 | 1 | 3 |
| Tab.[b] | $\begin{bmatrix} h_x & 1s_x \\ h_a & 1s_a \\ h_b & 1s_b \\ 2p_z \end{bmatrix}$ | $\begin{bmatrix} h_x & 1s_a \\ 1s_x & 1s_b \\ h_a & 2p_z \\ h_b \end{bmatrix}$ | $\begin{bmatrix} h_x & 1s_x \\ h_a & 1s_b \\ 1s_a & 2p_z \\ h_b \end{bmatrix}$ | $\begin{bmatrix} h_x & h_x \\ h_a & 1s_a \\ h_b & 1s_b \\ 2p_z \end{bmatrix}$ |
| $C_i(min)$ | 0.521 922 72 | −0.320 281 42 | −0.211 957 78 | 0.175 124 90 |

[a] The number of terms in the symmetry function that is generated from the tableau shown. (See text.)
[b] These tableau symbols exclude the core orbitals.

Table 13.6. *Second moments of the charge for CH$_3$.*

| Component | Value (D Å)[a] |
|---|---|
| $(xx + yy + zz)$ | −24.160 72 |
| $(2zz - xx - yy)/2$ | −1.829 70 |
| $(xx - yy)/2$ | 0.0 |
| $xz$ | 0.0 |
| $yz$ | 0.0 |
| $xy$ | 0.0 |

[a] Units of debye ångstroms.

Returning to the entries in Table 13.5, we see that the principal standard tableaux function is based upon the C $^5S$ state in line with our general expectations for this molecule with three C—H bonds. We considered in some detail the invariance of this sort of standard tableaux function to hybrid angle in our CH$_2$ discussion. We do not repeat such an analysis here, but the same results would occur. As we have seen in Chapter 6, standard tableaux functions frequently are not simply related to functions of definite spatial symmetry. The second and third standard tableaux functions are members of the same constellation as the first, but are part of pure $^2A_2''$ functions only when combined with other standard tableaux functions with smaller coefficients that do not show at the level to appear in the table. These other standard tableaux functions are associated with $LS$-coupled valence states of carbon at higher energies than that of $^5S$. The fourth term is ionic and associated with a negative C atom and partly positive H atoms.

The dipole moment of CH$_3$ is zero, of course, but the second moments of the charge have been determined, and $\langle r^2 \rangle$ and the quadrupole moments are given in Table 13.6. The sign of the $z$-axial quadrupole term indicates the distribution of

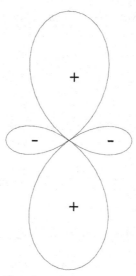

Figure 13.2. Drawing of positive axial quadrupole.

charge in the wave function. Figure 13.2 shows the general shape of the $z$-axial quadrupole with the signs of the regions. Since the moment of the molecule is negative, we see that its signs are reversed compared to those in the figure, and the individual C—H bonds are relatively positive at the H-atom ends.

We may arrive at this conclusion another way. In Table 13.6 the components $(xx - yy)/2$, $xy$, $xz$, and $yz$ are zero indicating that the quadrupole field is cylindrically symmetric about the $z$-axis. The *axial* moment around the $x$- or $y$-axis is

$$(2xx - yy - zz)/2 = 3(xx - yy)/4 - (2zz - xx - yy)/4, \quad (13.6)$$

$$= (2yy - xx - zz)/2, \quad (13.7)$$

$$= 0.91485 \, \mathrm{D\,\AA}, \quad (13.8)$$

for which the positive sign again indicates the positive nature of the H end of the C—H bonds. This direction of the dipole moment is the same as that of CH and $CH_2$, and is again expected because of the relative predominance in the wave function of the ionic term shown in Table 13.5.

## $CH_4$

We expect methane to be formed by the combination of an H-atom with the remaining unpaired $p_z$ orbital of $CH_3$. If the principal configuration is still that involving the $C^5S$ state and its nondirectional character predominates, we expect methane to be tetrahedral, thereby minimizing the repulsion energy between pairs of H atoms. This is borne out by the calculations as we see in Table 13.7.

Table 13.7. *Principal standard tableaux function structures for* $CH_4$
*at equilibrium bond distances.*

| | 1 | 2 | 3 | 4 |
|---|---|---|---|---|
| Num.[a] | 1 | 1 | 4 | 1 |
| Tab.[b] | $\begin{bmatrix} h_a & 1s_a \\ h_b & 1s_b \\ h_c & 1s_c \\ h_d & 1s_d \end{bmatrix}$ | $\begin{bmatrix} h_a & 1s_b \\ 1s_a & 1s_c \\ h_b & h_d \\ h_c & 1s_d \end{bmatrix}$ | $\begin{bmatrix} h_d & h_d \\ h_a & 1s_a \\ h_b & 1s_b \\ h_c & 1s_c \end{bmatrix}$ | $\begin{bmatrix} h_a & 1s_a \\ h_b & 1s_c \\ 1s_b & h_d \\ h_c & 1s_d \end{bmatrix}$ |
| $C_i(min)$ | 0.372 037 96 | −0.155 900 88 | 0.129 792 00 | −0.105 365 69 |

[a] The number of terms in the symmetry function that is generated from the tableau shown. (See text.)
[b] These tableau symbols exclude the core orbitals.

Table 13.8. *Apparent partial electronic charge on H atoms based upon lowest nonzero moment and the corresponding calculated bond lengths, STO3G basis.*

| Molecule | Charge | $R_{CH}$Å |
|---|---|---|
| CH | 0.254$\|e\|$ | 1.231 |
| $CH_2$ | 0.137 | 1.163[a] |
| $CH_3$ | 0.194 | 1.145 |
| $CH_4$ | 0.188 | 1.150 |

[a] The H—C—H bond angle is 129.1°.

A full valence calculation on $CH_4$ gives 1764 standard tableaux functions, and all of these are involved in the 164 $^1A_1$ symmetry functions. The second and fourth tableaux are also present in the principal constellation and, as with the earlier cases, these are not simple symmetry functions alone. The third tableau is ionic with the negative charge at the C atom. As before, this contributes to the relative polarity of the C—H bonds.

This is seen from a calculation of the electric moments. Methane has no nonzero dipole or quadrupole moments, but the $xyz$ component of the octopole is 1.144 D Å$^2$. All of the others are zero if the molecule has the orientation in the coordinate system that is used here. The value is positive, showing the same qualitative electronic distribution in C—H bonds as was seen for the other $CH_n$ molecules we have examined. Quantitatively, the octopole moment is equivalent to a charge of 0.204$\|e\|$ at the H-atom nuclei.

For easy comparison we show in Table 13.8 the apparent charge on the H atom in each of our molecules. The trend in these charges is broken between one and

Table 13.9. *Statistics for 6-31G\* calculations of* $CH_n$.

|  | State | Num. symm. funcs. | Number of tableaux |
|---|---|---|---|
| CH | $^2\Pi$ | 213 | 546 |
| $CH_2$ | $^3B_1$ | 828 | 1651 |
| $CH_3$ | $^2A_2''$ | 1597 | 9375 |
| $CH_4$ | $^1A_1$ | 2245 | 26 046 |

two H atoms. The likely interpretation here is that this is the place where the most important atomic configuration changes as one progresses through the list. This is seen clearly in Tables 13.2, 13.3, 13.5, and 13.7, where the principal configuration in the wave functions is shown.

In the early days of VB theory workers were concerned with the "valence state" of carbon[55]. Our calculations cannot really address this question because it is well defined only within a perfect pairing single tableau wave function.[3] The notion was contrived to explain the relatively constant bond energies through the $CH_n$ series, while there is a requirement to pay back the energy loss in having the principal configuration change to higher energy. In the context of a full valence calculation we may only give a somewhat more qualitative argument. The $^5S$ state of C is about 4 eV above the ground state. This suggests that each of the actual C—H bond energies in $CH_2$ with respect to some hypothetical frozen carbon state is about 2 eV higher than the apparent calculated or measured value. We attribute this to the greater effectiveness for bonding when $sp^n$ hybrids are involved.

### 13.1.2 6-31G\* basis

After our discussion of the STO3G results we, in this section, compare some of these obtained with a 6-31G\* basis arranged as described in Chapter 9. As before, we find that the larger basis gives more accurate results, but the minimal basis yields more useful qualitative information concerning the states of the atoms involved and the bonding. The statistics on the number of symmetry functions and standard tableaux functions for the various calculations are given in Table 13.9.

From Table 13.10 we see that the bond distances are reproduced better in this case than with the STO3G basis. We see that the break in the trend between CH and $CH_2$ again appears, and we continue to attribute it to the change in the important atomic configuration at this juncture in the list. The calculated bond distances are about 4.2% high. The success in calculating bond energies is more difficult to assess, since there is considerably more uncertainty in the experimental results.

---

[3] Even then, it is a purely theoretical concept. There appears to be no experimental approach to the energy of this state.

Table 13.10. *Energies, bond distances, and bond energies of* $CH_n$
*for 6-31G\* bases.*

| | | Bond length (Å) | | | $D_e$ (eV) | |
|---|---|---|---|---|---|---|
| | Energy (au) | Calc. | Exp. | Dissociation | Calc. | Exp. |
| C | −37.712 51 | | | | | |
| CH | −38.321 54 | 1.169 | 1.1190 | $D_{C-H}$ | 2.978 | 3.462 |
| $CH_2$ | −38.983 55 | $1.104^a$ | 1.029 | $D_{CH-H}$ | 4.502 | 4.33 |
| $CH_3$ | −39.624 08 | 1.109 | 1.079 | $D_{CH_2-H}$ | 3.918 | ≤4.90 |
| $CH_4$ | −40.295 47 | 1.119 | 1.094 | $D_{CH_3-H}$ | 4.758 | 4.406 |

$^a$ The H—C—H bond angle is 130.5°.

Table 13.11. *Various multipole moments and the apparent
charges on H atoms from 6-31G\* calculations.*

| | Moment$^a$ | Value | Charge |
|---|---|---|---|
| CH | D | 1.20030 | 0.214$|e|$ |
| $CH_2$ | D | 0.53033 | 0.119 |
| $CH_3$ | AQ | −1.33895 | 0.152 |
| $CH_4$ | O | 0.58764 | 0.022 |

$^a$ D, dipole; AQ, axial quadrupole; O, octopole.

The apparent charges on the H atoms in this basis are shown in Table 13.11. These may be compared to the similar values in Table 13.8. We see that the larger basis yields smaller values, particularly for methane. Nevertheless, we still predict that the H atoms in these small hydrocarbons are more positive than the C atom.

## 13.2 Ethane

Ethane presents a considerably greater challenge for calculation than the single carbon molecules above. Even if we continue the practice of putting $1s$ electrons in the "core" we have seven bonds and 14 electrons. A full minimal basis calculation, such as with STO3Gs, will produce 2 760 615 standard tableaux functions or HLSP functions for the total 14-electron basis. Not all of these are $^1A_{1g}$ (assuming $D_{3d}$ symmetry) but the number would be considerable. With 14 electrons and 14 orbitals, none doubly occupied, there are 429 possible Rumer diagrams or standard tableaux functions. We will not attempt any "full" calculations with ethane, but rather focus on basis set arrangements that are designed to yield useful results with greater efficiency.

Table 13.12. *Energies for covalent only calculations of $D_{3d}$ and $D_{3h}$ ethane.*

| Treatment | Num. symm. funcs. | Energy (hartree) | |
|---|---|---|---|
| | | $D_{3d}$ | $D_{3h}$ |
| Cartesian AO | 52 | −78.367 895 | [a] |
| Hybrid AO | 52 | −78.577 391 | −78.575 229 |
| Perfect pairing (hybrid) | 1 | −78.565 885 | −78.563 937 |

[a] This was not run.

We first contrast using a Cartesian basis with $sp^3$ hybrids on the C atoms for a covalent-only calculation.[4] Table 13.12 shows these along with the perfect pairing energy. We see that there is a considerable lowering of the energy at $\Delta E = 5.7$ eV from using hybrid orbitals on the C atom instead of the original Cartesian basis. The hybrids are arranged to be pointing at the H atoms and the other C atom. We also see that the perfect pairing wave function is not a great deal higher in energy than the full covalent-only energy at $\Delta E = 0.313$ or 0.307 eV for the $D_{3d}$ or $D_{3h}$ geometry, respectively. The perfect pairing function is the only Rumer tableau that is a symmetry function by itself. We saw earlier that a perfect pairing function with Cartesian AOs is frequently not sensible, and this is another such case.

Because they have no ionic states, the previous covalent-only results have too high a kinetic energy contribution, as discussed in Chapter 2. Adding all possible ionic states would lead to the very large number of basis functions quoted in the first paragraph of the discussion of ethane. We will consider the following physical arguments that may be used to limit the number of ionic state functions. This will all be done in the context of hybrid orbitals on the C atoms.

1. Adjacent ionic structures are the most important. This is expected since reductions in the kinetic energy will only occur if the overlap between the orbitals is fairly sizable. This is accomplished by assigning two electrons to each pair of orbitals that are arranged to bond in the molecule, and then requiring that this pair always have two electrons occupying them.
2. Only a few ionic bonds are required. We accomplish this by restricting the number of doubly occupied orbitals in a structure.
3. Highly charged atoms are unlikely. We accomplish this by preventing the charge depletion or build-up on either C atom from being outside $\pm 1$.

Table 13.13 shows the energies for several treatments of ethane using these arguments. The first addition of one set of ionic structures per basis function produces

[4] The reader is reminded that different linear combinations of the AOs yield different energies for less than full treatments.

Table 13.13. *Energies for various hybrid orbital calculations of $D_{3d}$ and $D_{3h}$ ethane.*

| Ionic structures | Num. symm. funcs. | Tableaux | Energy (hartree) | |
|---|---|---|---|---|
| | | | $D_{3d}$ | $D_{3h}$ |
| 0 | 52 | 429 | −78.577 391 | −78.575 228 |
| 1 | 214 | 2277 | −78.731 700 | −78.730 171 |
| 2 | 448 | 4797 | −78.742 547 | −78.741 195 |

Table 13.14. *Internal rotation barrier in ethane.*

| Ionic structures | Energy (eV) | |
|---|---|---|
| | Theory | Exp. |
| 0 | 0.059 | |
| 1 | 0.042 | |
| 2 | 0.037 | |
| | | 0.127[a] |

[a] See Ref. [56].

a lowering of $\approx$4.2 eV, or nearly 0.6 eV per bond. The second ionic structure produces only 0.04 eV more per bond. In Chapter 2 the lowering of the energy in $H_2$ when the ionic states are added is nearly 1 eV. The overlap there is rather greater at $\approx$0.9 than the values here, which are around 0.7 for either a C—H or a C—C bond.

We have calculated ethane in both $D_{3d}$ and $D_{3h}$ geometries. From Table 13.13 we obtain the calculated barriers to internal rotation given in Table 13.14. It is seen that the calculated barrier height is falling as the number of ionic states increases. It is not yet converged, but we do not give the result obtained by including three ionic structures in the basis functions. The interested reader can work this out. The trend here with the addition of ionic states runs counter to predictions using another method published by Pophristic and Goodman[57].

In addition it appears that this minimal basis calculation is unable to give a result close to the experimental value for the rotation barrier. We do not pursue this further here, but leave it as an open question.

## 13.3 Conclusions

In its original form VB theory was proposed using only states of atoms like the $^5S$ for C that we have invoked in describing our results. These are produced by standard tableaux functions of the particular sort that is antisymmetric with respect to the interchange of any of the four C orbitals $2s$, $2p_x$, $2p_y$, and $2p_z$. The functions based upon the other standard tableaux of the constellation correspond to the inclusion of other $LS$ states of the same configuration. Although not as important in the wave function, these functions do enter and allow one to infer that the step suggested by Slater and Pauling, the inclusion of all states of a configuration, was an important addition to the VB method.

Our results in this chapter also show that using hybrid orbitals with restricted bases can make an important improvement in the wave functions, at least when the criterion is energy lowering.

We also see that the number of basis functions grows rapidly with the number of electrons. In Chapter 16 we will discuss another method for dealing with the escalation of basis size with greater numbers of atoms and electrons.

# 14

## Rings of hydrogen atoms

In this chapter we examine some results for four model systems consisting of rings of H atoms. These calculations show how the number of atoms in a complex reaction may influence rates of reaction, particularly through the activation energy. The systems are as follows.

- Four H atoms in a rectangular geometry of $D_{2h}$ symmetry. The rectangle is characterized by two distances, $RA$ and $RB$. We map out a region of the ground state energy for this four-electron system as a function of the two distances.
- Six H atoms in a hexagonal geometry of $D_{3h}$ symmetry. This is not a *regular* hexagon, in general, but, like the system of four H atoms, is characterized by two distances we also label $RA$ and $RB$. These two distances alternate around the ring. We also calculate the map of the ground state energy for this six-electron system.
- Eight H atoms in an octagonal geometry of $D_{4h}$ symmetry and the specific shape characterized by the $RA$ and $RB$ variables as above. For this larger system we only determine the saddle point with respect to the same sort of variables.
- Ten H atoms in a decagonal geometry of $D_{5h}$ symmetry and the $RA$ and $RB$ variables. Again, we determine only the saddle point.

Since the geometries of these systems are in most regions not regular polygons, we will symbolize them as $(H_2)_n$, emphasizing the number of $H_2$ molecules rather than the total number of atoms.

For any of these, if $RA = 0.7$ Å and $RB$ is quite large, the rings represent 2–5 normal $H_2$ molecules well separated from one another. If the roles of $RA$ and $RB$ are reversed, the $H_2$ molecules have executed a metathesis in which the molecules transform into an equivalent set.

These are, without doubt, somewhat artificial systems. For real systems, one could not tell if anything happened, unless isotopic labeling could be arranged. An even greater problem would occur in the gas phase, since the entropy penalty required for these peculiar geometries would be expected to make them very improbable.

Table 14.1. *Number of symmetry functions of three types for H-ring calculations of $(H_2)_2$ and $(H_2)_3$.*

|        | Base configs. | Single exc. | Double exc. | Total | State |
|--------|---------------|-------------|-------------|-------|-------|
| $(H_2)_2$ | 8           | 17          | 33          | 58    | $^1A_{1g}$ |
| $(H_2)_3$ | 13          | 130         | 411         | 554   | $^1A_1'$ |

Table 14.2. *Number of symmetry functions for saddle point calculations of $(H_2)_4$ and $(H_2)_2$.*

|        | Num. Symm. Funcs. | Num. tab. | State |
|--------|-------------------|-----------|-------|
| $(H_2)_4$ | 146            | 1134      | $^1A_{1g}$ |
| $(H_2)_5$ | 768            | 7602      | $^1A_1'$ |

Nevertheless, the results have considerable interest, bearing, as they do, on the same sort of considerations as the Woodward–Hoffman rules[58].

## 14.1 Basis set

The calculations were all performed with an "$s$"-only basis of a $1s$ and a "$2s$" at each center. These are written in terms of the Huzinaga 6-Gaussian function as (6/42). This is the $s$ part of the basis used in Chapter 2 for the $H_2$ molecule and is shown in Table 2.2. It will be recalled that the "$2s$" orbital is not a real H2s orbital, but the second eigenfunction for this basis. As such it provides orbital breathing flexibility in the wave function. We show some statistics for these calculations in Table 14.1. Ionic states are restricted to $\pm1$ at any center. The saddle point calculations for the larger two systems were carried out with more restricted bases involving valence-only covalent and single-, and double-ionic structures. The statistics for these are shown in Table 14.2.

## 14.2 Energy surfaces

The energy surface for $(H_2)_2$, divided by 2, is shown in Fig. 14.1, and that for $(H_2)_3$, divided by 3, is in Fig. 14.2. Because of the division by the number of $H_2$ molecules, the energy goes to $-1$ hartree as $RA$ and $RB$ both grow large. Examination of the two surfaces shows clearly that they are quite different. The $(H_2)_2$ energy surface has a fairly sharp ridge between the two stable valleys. This is completely missing in the $(H_2)_3$ case. The difference between the energies

$$E_{H_4}/2 - E_{H_6}/3$$

Energy (hartree)

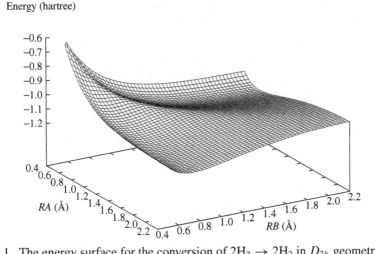

Figure 14.1. The energy surface for the conversion of $2H_2 \rightarrow 2H_2$ in $D_{2h}$ geometries. The energy is per $H_2$ molecule.

Energy (hartree)

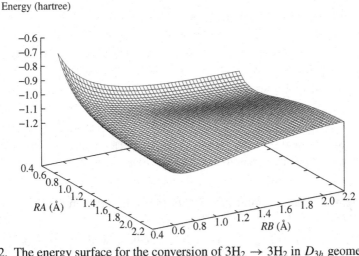

Figure 14.2. The energy surface for the conversion of $3H_2 \rightarrow 3H_2$ in $D_{3h}$ geometries. The energy is per $H_2$ molecule.

for the two surfaces is plotted in Fig. 14.3. This is everywhere $>-0.001$ eV within the region of the plot. Thus $E_{H_4}/2$ is always relatively higher.

In Table 14.3 we show the saddle points and activation energies of the four systems. It is seen that there is a tendency for the quantities to alternate between higher and lower values as the number of $H_2$ molecules is either even or odd. The differences decrease, however, as the rings become larger, and it appears that further calculations might show that the effect levels out. Nevertheless, the activation energy for the $(H_2)_2$ system is almost three times higher than that of the

Table 14.3. *Properties of the saddle points for the four hydrogen rings.*

|           | $RA$ Å | $RB$ Å | Energy au | Activation energy eV |
|-----------|--------|--------|-----------|----------------------|
| $(H_2)_2$ | 1.310  | 1.310  | −1.0367   | 3.04                 |
| $(H_2)_3$ | 0.998  | 0.998  | −1.1067   | 1.14                 |
| $(H_2)_4$ | 1.203  | 1.203  | −1.0475   | 2.75                 |
| $(H_2)_5$ | 1.107  | 1.107  | −1.0819   | 1.81                 |

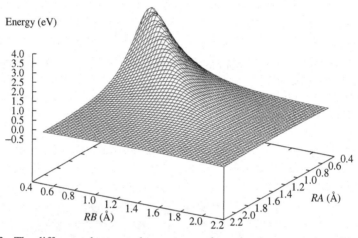

Figure 14.3. The difference between the energy surfaces for the conversion of $2H_2 \rightarrow 2H_2$ in $D_{2h}$ geometries and $3H_2 \rightarrow 3H_2$ in $D_{3h}$ geometries. The energy is per $H_2$ molecule. NB We have changed the aspect from that of Figs. 14.1 and 14.2.

$(H_2)_3$ system, the saddle point is 0.3 Å farther out and therefore more difficult to reach.

We can gain further insight into the differences between these two systems from an examination of the two $2 \times 2$ matrix systems based upon the Kekulé-like HLSP functions. These are, of course, the only structures for $(H_2)_2$, but for this comparison we ignore the long-bond functions in the other rings. We show the elements of these two matrix systems in Table 14.4. These systems are particularly simple since the diagonal elements of $H$ are equal, giving simple expressions for the eigenvalues of the problem, *viz.*,

$$E = \frac{H_{11} \pm H_{12}}{1 \pm S_{12}}, \tag{14.1}$$

$$= H_{11} \pm \frac{H_{12} - S_{12}H_{11}}{1 \pm S_{12}}. \tag{14.2}$$

Table 14.4. *Values of elements in the two-state HLSP function matrix systems for four hydrogen rings, all evaluated at the saddle points.*

| | $2H_{11}/n(=2H_{22}/n)$ | $2H_{12}/n$ | $S_{12}$ | $2(H_{12}-S_{12}H_{11})/n$ |
|---|---|---|---|---|
| $(H_2)_2$ | $-1.0113$ | $0.7058$ | $-0.6763$ | $0.02196$ |
| $(H_2)_3$ | $-1.0279$ | $-0.6308$ | $0.5752$ | $-0.03955$ |
| $(H_2)_4$ | $-1.0422$ | $0.3284$ | $-0.3015$ | $0.01418$ |
| $(H_2)_5$ | $-1.0452$ | $-0.2474$ | $0.2242$ | $-0.01307$ |

As we see from Table 14.4, there is a peculiar alternation in sign in passing through the series. This results in the lower energy arising from the upper or lower sign in Eq. (14.2) for $n$ odd or even in $(H_2)_n$, respectively. If we represent the two Kekulé structure functions by the symbols $K_1^{(n)}$ and $K_2^{(n)}$, the wave function at the saddle point is

$$\Psi_{\text{sad}}^{(n)} = N_n\left[K_1^{(n)} - (-1)^n K_2^{(n)}\right]. \tag{14.3}$$

The immediate consequence of this is a tendency for the electrons to stay away from the center of the ring for the even $n$ systems. For example, consider the $n = 2$ case:

$$\Psi_{\text{sad}}^{(2)} = 0.54615\left(\begin{bmatrix} a & b \\ c & d \end{bmatrix}_R - \begin{bmatrix} b & c \\ d & a \end{bmatrix}_R\right), \tag{14.4}$$

where $a$, $b$, $c$, and $d$ are the four orbitals around the ring, in order. This function would certainly be zero if the electrons were at locations such that all of the orbitals were of equal value. Because of the Pauli principle each of the Rumer tableaux functions is also zero at such a point, but there is an extra tendency toward zero because of the difference in Eq. (14.4). These do not occur in the odd $n$ cases. Such a point is the center of the ring. Therefore, we interpret the in-and-out alternation of the saddle point as a result of the extra tendency of the electrons in the even systems to avoid the center. As the ring becomes larger and the center farther away, the effect would be expected to decrease.

We also note that the values of the energy and overlap elements vary monotonically with $n$, contrary to the alternating characteristic so far emphasized. Their specific values give a maximum at $n = 3$ in the last column, however. The maximum remains after division by $(1 \pm S_{12})$. Thus the $(H_2)_3$ system has the largest interaction at the saddle point, as measured by the energy decrease when the two structures interact. It is also interesting that $H_{11}$ decreases as $n$ increases. An immediate explanation for this is not available.

The real reactions that most resemble these are the production of cyclobutane from two ethylene molecules $((H_2)_2)$ and the Diels–Alder reaction between butadiene and ethylene $((H_2)_3)$. Even these cannot be made to react in the bare forms, but fairly simple activation by substituents will allow the $(H_2)_3$ analog to proceed. Apparently, no form of the $(H_2)_2$ analog has ever been observed. Our analysis suggests that there is a fundamental difference between the four-electron and six-electron systems that produces the effect. The book by Woodward and Hoffman[58] may be consulted for a rationalization of these results based upon MO theory.

As we continue to larger rings, the results are not so clearcut. There is a tendency for the saddle points to alternate in and out somewhat, but the interaction energy appears to be a maximum at the ring of six atoms. These last more-or-less qualitative comments have been based upon just the two simple Kekulé-like structures of the rings, and may not be able to show the proper behavior. The actual surface calculations of $(H_2)_2$ and $(H_2)_3$ included many more structures and show strikingly different qualitative behavior.

# 15

# Aromatic compounds

Benzene is the archetypal aromatic hydrocarbon and its study has been central to the understanding of aromaticity and resonance from the early times. In addition, it has the physical property of having its $\pi$ electrons reasonably independent from those in $\sigma$ bonds, leading early quantum mechanics workers to treat the $\pi$ electrons alone. Since benzene is a ring and the rules for forming Rumer diagrams have one draw noncrossing lines between orbital symbols written in a circle, the Rumer diagrams correspond to the classical Kekulé and Dewar bond schemes that chemists had postulated far earlier than the VB treatments occurred. This parallel has intrigued people since its first observation and led to many discussions concerning its significance. It has also led to considerable work in more qualitative "graphical methods" for which the reader is directed to the literature. (See, *inter alia*, Randić[59].)

We will examine benzene with different bases and also discuss some of the ideas that consideration of this molecule has led to, such as resonance and resonance energy.

We show again the traditional five covalent Rumer diagrams for six electrons and six orbitals in a singlet coupling and emphasize that the similarity between the ring of orbitals and the shape of the molecule considerably simplifies the understanding of the symmetry for benzene.

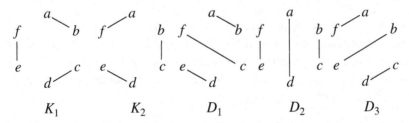

Most of the discussion we give here on the nature of the wave function will focus on HLSP functions. An early *ab initio* study by Norbeck and the present author

Table 15.1. *The four most important categories of HLSP functions in the wave function for benzene.*

| | 1 | 2 | 3 | 4 |
|---|---|---|---|---|
| Num.[a] | 2 | 3 | 12 | 12 |
| HLSP[b] | $\begin{bmatrix} 2p_a & 2p_b \\ 2p_c & 2p_d \\ 2p_e & 2p_f \end{bmatrix}_R$ | $\begin{bmatrix} 2p_b & 2p_c \\ 2p_a & 2p_d \\ 2p_e & 2p_f \end{bmatrix}_R$ | $\begin{bmatrix} 2p_a & 2p_a \\ 2p_c & 2p_d \\ 2p_e & 2p_f \end{bmatrix}_R$ | $\begin{bmatrix} 2p_a & 2p_a \\ 2p_c & 2p_c \\ 2p_e & 2p_f \end{bmatrix}_R$ |
| $C_i{}^c$ | 0.160 88 | −0.057 63 | 0.051 62 | 0.027 44 |

[a] The number of terms in the symmetry function that is generated from the tableau shown. (See text.)

[b] These tableau symbols exclude the core orbitals.

[c] In this case all of the terms in a symmetry function have the same sign as well as magnitude for the coefficient.

on benzene[60] focused more on interpretation of the standard tableaux function representation of the wave functions. Thus, the present discussion now differs from that earlier in some respects.

## 15.1 STO3G calculation

The Weyl dimension formula (Eq. (5.115)) tells us that six electrons in six orbitals in a singlet state yield 175 basis functions. These may be combined into 22 $^1A_{1g}$ symmetry functions. Table 15.1 shows the important HLSP functions for a $\pi$-only calculation of benzene for the SCF optimum geometry in the same basis. The $\sigma$ orbitals are all treated in the "core", as described in Chapter 9, and the $\pi$ electrons are subjected to its SEP. We discuss the nature of this potential farther in the next section. The functions numbered in the first row of Table 15.1 have the following characteristics.

1. The two functions of this type are the classical Kekulé structures for benzene. One might expect the coefficient to be larger, but we will see below why it is not.
2. These three functions are the classical Dewar structures.
3. The third set of functions, 12 in number, are all of the possible singly ionized structures where the charges are adjacent, and there are no long bonds.
4. The fourth set of functions are all of the doubly charged structures with the + and − charges adjacent and no long bonds.

When there is a relatively high degree of symmetry as in benzene, the interpretation of the parts of the wave function must be carried out with some care. This arises from an apparent enhancement of the magnitude of the coefficient of a structure in the wave function when whole symmetry functions are used. Let us consider the

Kekulé structures and denote them by $K_1$ and $K_2$. As discussed in Chapter 5, when these HLSP functions are projected from the appropriate product of $p$ orbitals

$$K_1 = \theta' \mathcal{PNP} \rho_1,$$
$$K_2 = \theta' \mathcal{PNP} \rho_2,$$

they are not normalized to 1. The "raw" $2 \times 2$ overlap matrix is

$$\begin{bmatrix} 2.246\,364\,9 & \\ 0.799\,515\,5 & 2.246\,364\,9 \end{bmatrix},$$

and, hence, the true overlap between a normalized $K_1$ and $K_2$ is $0.355\,915\,2$. Thus, if we consider the $^1A_{1g}$ *symmetry* function involving $K_1$ and $K_2$, we obtain in its normalized form

$$^1A_{1g} = 0.607\,251\,7(K_1 + K_2),$$

and, if the wave function is written in terms of this symmetry function, its coefficient would be $0.264\,931\,3$ instead of the number listed in Table 15.1 for the *individual* Kekulé structures. In these terms, the Kekulé structures appear to have a larger coefficient. A similar analysis for the Dewar structures leads to an apparent enhancement of the coefficient magnitude to $-0.133\,825\,9$.

The apparent enhancement we are discussing here is more pronounced, in general, the greater the number of terms in the symmetry function. We now consider the third sort of function from Table 15.1. These are the 12 short-bond singly ionic functions, and in this case the enhancement of the coefficient is a factor of 5.0685, i.e., the reciprocal of the normalization constant for the symmetry function that is the sum of the individually normalized HLSP functions. The resulting coefficient would then be $0.261\,637$, a number essentially the same as the coefficient of the Kekulé symmetry function.

Are the Kekulé functions and the short-bond singly ionic functions really of nearly equal importance in the wave function? This appears to be the only possible conclusion and may be rationalized as follows. We have seen that the covalent-only structures provide for a considerable electron correlation, lowering their potential energies, but constrain the space available to the electrons, thereby raising their kinetic energies. Ionic structures allow delocalization that lowers the kinetic energy while not raising the potential energy enough to prevent an overall decrease in energy. When there are six bonds to be delocalized we expect the effect in the singly ionic structures to be roughly six times that for only one bond. Those we discuss are the adjacent only ionic structures and are expected to be the most important.

We also observe that the diagonal element of the Hamiltonian for a single Dewar structure is about 1.8 eV higher than a diagonal element of a single Kekulé structure. This is a very reasonable number for the difference in energies between a long and a short bond. The short-bond singly ionic structures have a diagonal element of the

Hamiltonian that is about 8.6 eV above the single Kekulé structure. This number is not exactly comparable to the ionic structures in $H_2$ discussed in Chapter 2. Consider the two ionic structures,

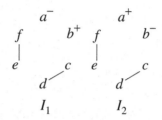

which are two of the 12 short-bond adjacent singly ionic structures in benzene. The $2 \times 2$ secular equation corresponding to these two functions is

$$\begin{vmatrix} 8.562 - E & E \\ -1.442 - 0.0995E & 8.562 - E \end{vmatrix} = 0,$$

where we have converted the energies to electron volts and have reset the zero to the energy of the single Kekulé structure, $K_1$. The lower root of this equation is 6.476 eV, which is $\approx 2$ eV lower than the energy of $I_1$ alone. We should compare this with the corresponding value for $H_2$, obtained with methods of Section 2.4, wherein 5.82 eV is obtained. Thus the effect on the diagonal energy of forming the ionic structure pair is in the same direction for the two systems, but much larger in the more compact $H_2$.

### 15.1.1 SCVB treatment of $\pi$ system

We have so far emphasized the nature of the wave function. We now examine the energies of some different arrangements of the bases. In Table 15.2 we show energies for five levels of calculation, Kekulé-only, Kekulé plus Dewar, SCF, SCVB, and full $\pi$ structures, where energies are given as the excess energy due to the $\pi$ system over that from the core. Cooper *et al.*[61] gave the SCVB treatment of benzene.

We note first that the covalent-only calculations give a higher energy than the SCF wave function. We noted this effect with the allyl radical in Chapter 10, and it happens again here with benzene. This is again a manifestation of the delocalization provided by ionic structures in the wave function and the concomitant decrease in the kinetic energy of the electrons. Since this phenomenon does not occur in cases where resonance is absent, we expect it to be greater where there are possibilities for greater numbers of more or less equivalent resonance structures.

There is only one equivalent orbital in a highly symmetric $\pi$ system like that in benzene. This is shown as an altitude plot in Fig. 15.1. We see that each orbital is

Table 15.2. *Comparison of different calculations*
*of the π system of benzene. All energies*
*are in hartrees.*

|  | $E - E_{core}{}^a$ |
|---|---|
| $K_1$ | −6.711 44 |
| $K_1$ & $K_2$ | −6.755 53 |
| Full cov. | −6.760 79 |
| SCF | −6.834 10 |
| SCVB | −6.904 88 |
| Full π | −6.911 87 |

$^a$ $E_{core} = -222.142\,48$.

Orbital amplitude

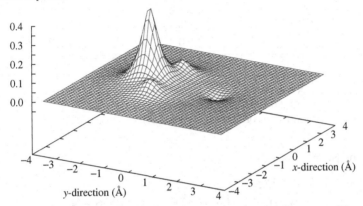

Figure 15.1. Altitude plot of the SCVB orbital for the π system of benzene. There are six symmetrically equivalent versions of this around the ring. The amplitude is given in a plane 0.5 Å in the positive $z$-direction from the plane of the nuclei.

predominantly a single $2p_z$ with smaller "satellites" in the ortho positions, essentially nothing at the meta positions, and a negative contribution at the para position.

The coefficients in the two covalent-only wave functions, $\Psi$(pure $2p_z$) and $\Psi$(SCVB $2p_z$) are not very different:

$$\Psi(\text{pure } 2p_z) = 0.402\,88(K_1 + K_2) - 0.150\,26(D_1 + D_2 + D_3),$$
$$\Psi(\text{SCVB } 2p_z) = 0.403\,53(K_1 + K_2) - 0.122\,03(D_1 + D_2 + D_3).$$

We may, however, examine the $2 \times 2$ Kekulé-only matrices for these two cases. For the pure $2p_z$ orbitals we have, in hartrees, the secular equation

$$\begin{vmatrix} -6.711\,44 - & E \\ -2.448\,49 - 0.355\,92E & -6.711\,44 - E \end{vmatrix} = 0,$$

Table 15.3. *Lowering of the energy from resonant mixing for pure and SCVB $2p_z$ orbitals.*

|              | Pure eV | SCVB eV |
|--------------|---------|---------|
| $K$ − only   | −1.199  | −0.870  |
| $K + D$      | −1.343  | −0.878  |

Table 15.4. *Comparison of some one- and two-electron matrix elements for pure and SCVB $2p_z$ orbitals. All energies are in hartrees.*

|            | Pure    | SCVB    |
|------------|---------|---------|
| $T^a$      | 1.2299  | 1.0891  |
| $V^b$      | −2.9613 | −2.8750 |
| $[11|11]^c$| 0.5795  | 0.5125  |
| $[11|22]^d$| 0.3242  | 0.3352  |

$^a$ Kinetic energy.
$^b$ Nuclear and core potential energy.
$^c$ Orbital self-repulsion energy.
$^d$ Adjacent orbital repulsion energy.

and for the SCVB orbitals,

$$\begin{vmatrix} -6.872\,56 - & E \\ -4.616\,57 - 0.664\,00E & -6.872\,56 - E \end{vmatrix} = 0.$$

We see that the $K_1$–$K_1$ diagonal element for the SCVB orbitals is already about 4.4 eV below that for the pure $2p_z$ orbitals. This is the most immediate explanation for the lower energy of the SCVB result. In fact, this is the larger effect. As seen in Eq. (14.2), the amount of energy lowering in $2 \times 2$ systems like these is

$$\frac{H_{12} - S_{12}H_{11}}{1 + S_{12}}.$$

In this case we have 1.20 eV and 0.87 eV for the pure $2p_z$ and SCVB orbitals, respectively, and the resonance appears somewhat more beneficial for the localized orbital. These results are included in Table 15.3.

We may obtain more information from a comparison of some of the one- and two-electron integrals for the individual orbitals. The values are shown in Table 15.4. It is seen that the changes in the potential energy terms nearly cancel with the repulsive self-energy and the nuclear and core potential energies changing in opposite directions. The change in the adjacent repulsion energy is also not

Table 15.5. *Comparison of different calculations of the π system of linear 1,3,5-hexatriene. All energies are in hartrees.*

|  | $E - E_{core}^a$ |
| --- | --- |
| $K_1$ | −6.129 74 |
| Full cov. | −6.133 78 |
| SCF | −6.164 40 |
| SCVB | −6.251 49 |
| Full π | −6.252 41 |

[a] $E_{core} = -223.91056$.

large. There remains only the kinetic energy term, for which the difference is nearly 3.8 eV, with the SCVB orbital lower. This is easily interpreted to be the result of the delocalization in that orbital, and it thus makes the same sort of contribution as do ionic structures in MCVB wave functions. The values of the $n$-electron matrix elements are the result of an interplay of considerable complexity among the simpler one- and two-electron matrix elements, and it is not really possible to say much more about the effects of these latter quantities upon the total energies.

We show another aspect of these numbers in Table 15.3, where we detail the effects of resonance between the two Kekulé structures and among all of the covalent structures for the two sorts of $2p_z$ orbitals. The results suggest that within one structure the SCVB orbitals duplicate, to some extent, the effect of multiple structures, and the configurational mixing produces less energy lowering with them.

Those familiar with the long history of the attacks on the question of the resonance energy of benzene may be somewhat surprised at the small numbers in Table 15.3. The energy differences that are given there are for just the sort of process that might be expected to yield a theoretical value for the resonance energy, but experimental determinations yield numbers in the range 1.7–2.3 eV. This is an important question, which we will take up in Section 15.3, where it will turn out that some subtleties must be dealt with.

### 15.1.2 Comparison with linear 1,3,5-hexatriene

In order to put the structure of benzene into better perspective, we give a similar calculation of 1,3,5-hexatriene for comparison. Table 15.5 shows the energies for a set of calculations parallel to those in Table 15.2 for benzene. The most obvious difference is the smaller total spread in the energies, about 3.3 eV rather than the 5.5 eV for benzene, and the SCVB energy is closer to the full π than in benzene. Standard arguments say that there is only a rather small amount of resonance

Orbital amplitude

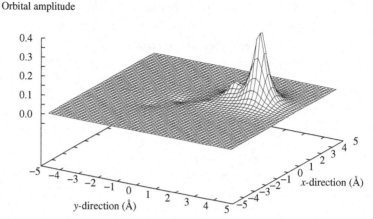

Figure 15.2. Altitude plot of the first SCVB orbital for the $\pi$ system of 1,3,5-hexatriene. There are two symmetrically equivalent versions of this at each end of the molecule. The amplitude is given in a plane 0.5 Å in the positive $z$-direction from the plane of the nuclei.

Orbital amplitude

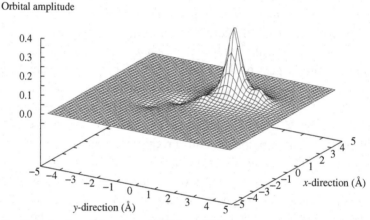

Figure 15.3. Altitude plot of the second SCVB orbital for the $\pi$ system of 1,3,5-hexatriene. There are two symmetrically equivalent versions of this at each end of the molecule. The amplitude is given in a plane 0.5 Å in the positive $z$-direction from the plane of the nuclei.

in linear hexatriene, since only the one Kekulé structure has only short bonds, nevertheless, the SCF energy is lower than the full covalent energy.

There are three inequivalent SCVB orbitals for hexatriene, and these are given in Figs. 15.2, 15.3, and 15.4. The first of these shows a principal peak at the first $2p_z$ orbital and a small satellite at the adjacent position. The second is more interesting with the principal peak at the second C atom, but showing a larger satellite at position 1 than at position 3. This is consistent with having essentially a double bond between atoms 1 and 2 or 5 and 6, with single bonds between 2 and 3 and 4 and 5.

Table 15.6. *Comparison of different calculations of*
*the $\pi$ system of benzene with a 6-31G\* basis.*
*All energies are in hartrees.*

|  | $E - E_{core}{}^a$ |
|---|---|
| SCF | $-6.416\,20$ |
| Full valence $\pi$ | $-6.464\,30$ |
| SCVB | $-6.479\,44$ |
| Full valence $\pi + S^b$ | $-6.496\,50$ |

$^a$ $E_{core} = -224.275\,51$.
$^b$ Single excitation.

Orbital amplitude

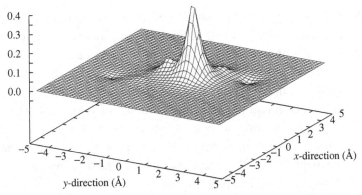

Figure 15.4. Altitude plot of the third SCVB orbital for the $\pi$ system of 1,3,5-hexatriene. There are two symmetrically equivalent versions of this at each end of the molecule. The amplitude is given in a plane 0.5 Å in the positive $z$-direction from the plane of the nuclei.

The third orbital has a similar interpretation with a larger satellite at position 4 than at position 2.

The overall conclusion is that there is considerably less resonance in hexatriene than in benzene, and the bond lengths and types alternate along the chain unlike the equivalence in benzene.

## 15.2 The 6-31G\* basis

The SCF, SCVB, full valence $\pi$, and full valence $\pi + S$ results of using a 6-31G\* basis on benzene are given in Table 15.6. The geometry used is that of the minimum SCF energy of the basis. In this case the SCVB energy is lower by 0.4 eV than the full valence $\pi$ energy. This is principally due to the $3d$ polarization orbitals present in the SCVB orbital, but absent in the valence calculation. The SCVB orbital is

Table 15.7. *The centroids of charge implied by the second moment of the charge distribution of the nuclear and $\sigma$ framework. The C and H nuclear positions are those of the 6-31G\* SCF equilibrium geometry.*

| Charge | Radial distance (Å) |
|--------|---------------------|
| $C^{+6}$ | 1.3862 |
| $-5\|e\|$ | 1.8730 |
| $H^{+1}$ | 2.4618 |

quantitatively so close to that from STO3G orbitals shown in Fig. 15.1 that the eye cannot detect any difference, and we do not draw a 6-31G\* version of the orbital.

The SEP is used again to represent the core, and the following analysis may be made to get a crude picture of its nature. The second moment of the $xx(=yy)$ charge distribution of the core is 0.481 10 bohr$^2$. At each apex of the hexagon there is a C nucleus and an H nucleus farther out. There are also five electrons per apex contributed by the $\sigma$ system. The (quadratic) centroid of this charge may be calculated from the second moment, and is shown in Table 15.7. The overlap and kinetic energy one-electron matrix elements of the $\pi$ AOs are unaffected by the SEP. In addition, our centroid picture does not include any of the exchange effects present.[1] The main point is that arguments using a nuclear charge effect of $+1$ for each $\pi$ AO may be too simplistic for many purposes.

We saw in Chapter 14 that a ring of six H atoms does not want to be in a regular hexagonal geometry at the minimum energy. A question concerning benzene arises then: Is benzene a regular hexagon because of or in spite of the resonance in the $\pi$ system? The previous calculations have all been done with the regular hexagon geometry forced on the molecule. We now relax that constraint to test the stability of the ring against distortion into an alternating bond length geometry. In Table 15.8 we show the values and first and second derivatives of the SCF, core, valence, and total energies with respect to two distortion directions in the molecule. The first direction we call "$a_{1g}$" is a symmetric breathing motion involving the change in length of only the C—C bonds. (The C—H bonds are unchanged.) The second we call "$b_{2u}$", and it involves an alternating increase and decrease in the lengths of the C—C bonds around the rings.[2] (C—C—C, C—C—H angles, and C—H distances are not changed.)

---

[1] These are not expected to be very large in a system like benzene where there is a natural symmetry-based orthogonality between the $\pi$ and $\sigma$ systems.

[2] These group species symbols correspond to the symmetry of vibrational normal modes of the same sort.

Table 15.8. *Energies and derivatives of energies relating to the stability of the regular hexagonal structure of benzene.*

|  |  | SCF | Core | $\pi$ | Total |
|---|---|---|---|---|---|
| Energy[a] |  | −230.691 71 | −224.275 51 | −6.496 50 | −230.772 02 |
| $\partial E/\partial(\Delta R)$[b] | $a_{1g}$ | −0.013 8 | −2.050 6 | 1.981 1 | −0.069 5 |
|  | $b_{2u}$ | 0.0 | 0.0 | 0.0 | 0.0 |
| $\partial^2 E/\partial(\Delta R)^2$[c] | $a_{1g}$ | 3.492 7 | 5.287 8 | −1.878 2 | 3.409 6 |
|  | $b_{2u}$ | 5.513 7 | 16.742 9 | −10.517 4 | 6.225 5 |

[a] Hartrees.
[b] Hartrees/bohr. $\Delta R$ is the change in the C—C distance in the ring in all cases.
[c] Hartrees/bohr².

At first glance the numbers in Table 15.8 suggest that the answer to the question in the last paragraph is the "in spite of" alternative. The last row of the table shows that the $b_{2u}$ distortion has a stable minimum in the core but a maximum in the $\pi$ energy. The values are such that the total has a stable minimum. The calculations thus correctly predict that there is no force tending to distort the regular hexagon at the point represented by the regular geometry.

Nevertheless, there is a difficulty with the interpretation in the last paragraph. There has developed over the years a considerable literature on this question, with many opinions on both sides. Shaik *et al.* have written articles on this subject[62, 63]. Such a situation frequently indicates the existence of ambiguities in the definition, and that certainly applies to this case. We may describe the situation according to our current terms.

As stated above, we have used the SEP to obtain separate energies for the $\sigma$ core and the $\pi$ system. Conventionally, this consists of attributing the whole of the nuclear repulsion energy to the core. We called attention, however, to the nature of the energy surface for six H atoms in a ring (see Chapter 14), which, unlike the benzene $\pi$ system in our treatment, does have nuclear repulsion included. It might be expected that we would need to make some sort of partitioning of the nuclear repulsion energy to make the H atom and the $\pi$ systems comparable. One way to do this would be to imagine the effect of one unit charge from the C nuclei being subtracted from the core energy and added to the $\pi$ energy. None of the totals is affected, of course. The result for the repulsion energy for six $|e|$ charges at the positions of the C atoms is 4.185 55 hartrees, but more important for our purposes is that the second derivative is 0.8067 hartrees/bohr² in terms of the same $\Delta R$ coordinate used in Table 15.8. It is seen that it would take the switch of many more charges from the core to the $\pi$ system than are actually present to reverse its tendency to distort the regular hexagon. Thus, we need not revise the earlier conclusion.

Table 15.9. *Energies and derivatives of energies relating
to the stability of the square geometry of cyclobutadiene.*

|                                      |         | Core       | $\pi$     | Total      |
|--------------------------------------|---------|------------|-----------|------------|
| Energy[a]                            |         | $-150.52261$ | $-3.08046$ | $-153.60307$ |
| $\partial E/\partial(\Delta R)$[b]   | $b_{2u}$ | 0.0        | 0.0       | 0.0        |
| $\partial^2 E/\partial(\Delta R)^2$[c] | $b_{2u}$ | 5.9223     | $-6.9506$  | $-0.7283$   |

[a] Hartrees.
[b] Hartrees/bohr.
[c] Hartrees/bohr$^2$.

The size of the ring is seen to be a compromise between attraction due to the
$\pi$ system and repulsion due to the core. Again, the sizes of the second derivatives
add to give a stable minimum. If we assume that the system behaves harmonically,
the derivatives in the last column imply that the ring C—C bond distances are about
0.0204 bohr (0.0108 Å) longer than the value used for the first row of the table,
which are the SCF minimum distances.

### 15.2.1 Comparison with cyclobutadiene

It is illuminating to compare the behavior of benzene[3] upon the $b_{2u}$ type of distortion
with a similar calculation in cyclobutadiene. We do not repeat all of the calculations
of the last section, but do include the results in Table 15.9, where we give the energies
and the derivatives for the $b_{2u}$ distortion from the geometry of a square. It will be
recalled from Chapter 14 that the energy surface for the ring of four H atoms has
a higher peak at the square geometry than the six-atom ring does at the regular
hexagon. The same situation applies here. The second derivative for the $\pi$ system
is larger in magnitude than the core derivative, making the square an unstable
structure for $C_4H_4$. Thus we predict that cyclobutadiene in a singlet state has no
tendency to form a molecule with equal ring bond lengths as happens in benzene.

In these discussions of benzene and cyclobutadiene we have compared MCVB
level calculations of the $\pi$ system with SCF level calculations of the core. We do
not expect that using correlated wave functions for core energies would change the
results enough to give a different qualitative picture.

### 15.3 The resonance energy of benzene

Once Kekulé had deduced the correct structure of benzene, chemists soon realized
that the double bonds in it were considerably more stable than isolated double

---

[3] The same species symbol can serve in $D_{4h}$ symmetry.

bonds in aliphatic hydrocarbons. The principal evidence is that reaction conditions leading to *addition* to an aliphatic double bond with the removal of the multiple bond do not normally affect benzene, and more vigorous conditions cause an attack that removes a ring H atom and leaves the double bonds unchanged. The conclusion was that the "conjugated" ring double bonds possess an added stability due to their environment. It remained for quantum mechanics to explain this effect in terms of what has come to be called "resonance" among a number of bond structures.

Experimental approaches to determining the resonance energy (called stabilization energy by some) have involved comparing thermodynamic measurements of benzene with those of three cyclohexenes. Heats of combustion and heats of hydrogenation have been used. Most feel the hydrogenation method to be superior, since it is expected to involve smaller differencing errors in the determination. The energies and processes are

$$C_6H_6 + 3H_2 \rightarrow C_6H_{12}; \Delta H = -2.13eV,$$
$$C_6H_{10} + H_2 \rightarrow C_6H_{12}; \Delta H = -1.23eV.$$

The difference,[4] $-1.54$ eV, corresponds to the lower energy the three double bonds in benzene have than if they were isolated. This is not much larger than the "pure $2p_z$" entry in the second row of Table 15.4. It was pointed out by Mulliken and Parr[64], however, that this precise comparison is not what should be done. The number in the table from our calculation does not involve any change in the bond lengths whereas the experiment certainly does. Changes in energy due to bond length change come from both the $\pi$ bonds and the $\sigma$ core.

It is possible to make a successful comparison of theory with experiment for the resonance energy modified according to the Mulliken and Parr prescription[60], but there are still many assumptions that must be made that have uncertain consequences. A better approach is to attempt calculations that match more closely what experiment gives directly. This still requires making calculations on what is a nonexistent molecule, but the unreality pertains only to geometry, not to restricted wave functions.

Following these ideas, Table 15.10 shows results of 6-31G* calculations of the $\pi$ system of normal benzene and benzene distorted to have alternating bond lengths matching standard double and single bonds, which we will call cyclohexatriene.

The cyclohexatriene molecule has a wave function considerably modified from that of benzene. The first few terms are shown in Table 15.11, where the two

---

[4] Most workers change the sign of this to make it positive, but logically an energy corresponding to greater stability should be negative.

Table 15.10. *Calculations of π energies for normal and distorted benzene.*

|                  | SCF        | MCVB           |
|------------------|------------|----------------|
| Benzene          | −6.416 20  | −6.496 50[a]   |
| Cyclohexatriene  | −6.363 07  | −6.442 91[b]   |
| $\Delta E$ eV    | −1.446     | −1.458         |

[a] This gives 3211 tableaux functions formed into 280 symmetry functions.

[b] This gives 3235 tableaux functions formed into 545 symmetry functions.

Table 15.11. *The first terms in the MCVB wave function for cyclohexatriene.*

|       | 1 | 2 | 3 | 4 |
|-------|---|---|---|---|
| Num.  | 1 | 6 | 1 | 3 |
| HLSP  | $\begin{bmatrix} 2p_a & 2p_b \\ 2p_c & 2p_d \\ 2p_e & 2p_f \end{bmatrix}_R$ | $\begin{bmatrix} 2p_a & 2p_a \\ 2p_c & 2p_d \\ 2p_e & 2p_f \end{bmatrix}_R$ | $\begin{bmatrix} 2p_b & 2p_c \\ 2p_d & 2p_e \\ 2p_a & 2p_f \end{bmatrix}_R$ | $\begin{bmatrix} 2p_c & 2p_d \\ 2p_b & 2p_e \\ 2p_a & 2p_f \end{bmatrix}_R$ |
| $C_i$ | 0.275 450 | 0.082 38 | 0.053 11 | −0.040 78 |

"Kekulé" structures have quite different coefficients. We interpret the terms as follows.

1. This is the standard Kekulé structure with the π bonds principally at the short distance.
2. The second group of functions, six in number, are adjacent single ionic structures corresponding to bonds in the position marked in function 1.
3. Function 3 is the other Kekulé structure. Its importance in the wave function is low, indicating little π bonding at the long positions.
4. The fourth group consists of the three "Dewar" structures and is also relatively unimportant.

When we look at the energies from Table 15.10, perhaps the most striking fact is that the correlation energy in the π system makes so little difference in the $\Delta E$ values. As we indicated above, the experimental value for the resonance energy from heats of hydrogenation is −1.54 eV, in quite satisfactory agreement with the result in Table 15.10. The fact that our value is a little lower than the experimental one may be attributed to the small amount of residual resonance remaining in the cyclohexatriene, whereas the isolated double bonds in the experiment are truly isolated in separate molecules.[5]

---

[5] There is still interest in the resonance energy of benzene. Beckhaus *et al.*[65] have synthesized a molecule with a strained benzene ring in it and measured heats of hydrogenation. This is an experimental attempt to assay what we did theoretically. They found similar results.

Table 15.12. *Core, π SCF, and π MCVB energies for various calculations of naphthalene. An STO3G basis is used, and all energies are in hartrees.*

|  |  | Energy | Num. symm. funcs. |
|---|---|---|---|
|  | Core | −366.093 70 |  |
|  | SCF $\pi$ | −14.398 73 |  |
| "Kekulé" | VB | −14.221 75 | 1 |
| Covalent | MCVB | −14.277 12 | 16 |
| Covalent + single-ionic | MCVB | −14.476 32 | 334 |
| Covalent + single-and double-ionic | MCVB | −14.524 33 | 1948 |
| Full $\pi$ | MCVB | −14.529 93 | 4936 |

## 15.4 Naphthalene with an STO3G basis

We now consider naphthalene, which possesses 42 covalent Rumer diagrams. Many of these, however, will have long bonds between the two rings and are probably not very important. To the author's knowledge no systematic *ab initio* study has been made of this question. The molecule has $D_{2h}$ symmetry, and these 42 covalent functions are combined into only 16 $^1A_g$ symmetry functions.

As with benzene we study only the $\pi$ system using the SEP to account for the presence of the $\sigma$ orbitals. It is not the purpose of this book to compare MCVB with molecular orbital configuration interaction (MOCI) results, but we do it in this case.

### 15.4.1 MCVB treatment

We first give the MCVB results in Table 15.12, which shows energies for several levels of calculations with an STO3G basis. A full $\pi$ calculation for naphthalene consists of 19 404 singlet tableau functions, which may be combined into 4936 $^1A_g$ symmetry functions. The covalent plus single ionic calculation involves 1302 singlet tableau functions, which may be combined into 334 symmetry functions, and the covalent, single-, and double-ionic treatment produces 7602 singlet tableau functions, which may be combined into 1948 symmetry functions.

The results in Table 15.12 show again that the SCF function has a lower energy than the covalent-only VB. Although a thorough study has not been made, it appears that this difference increases with the size of the system. Certainly, the decrease in energy upon adding the single-ionic structures to the basis is greater here than in benzene, <4.1 eV for benzene versus 5.42 eV for naphthalene. Again we see the

Table 15.13. *Energies for MOCI* $\pi$ *-only calculations of*
*naphthalene for different levels of excitation.*

|  | Energy | Num. symm. funcs. |
|---|---|---|
| Core | −366.093 70 | |
| SCF | −14.398 73 | 1 |
| Single | −14.398 73 | 7 |
| Double | −14.512 40 | 98 |
| Triple | −14.514 83 | 522 |
| Quadruple | −14.528 82 | 1694 |
| Full | −14.529 93 | 4936 |

importance of delocalization in the wave function. The full delocalization energy provided by including all ionic structures is 6.88 eV compared with 4.11 eV for benzene (see Table 15.2). The ratio here is 1.67, remarkably close to the ratio of the numbers of electrons in the two $\pi$ systems. In contrast, the delocalization energy in 1,3,5-hexatriene is only 3.23 eV (see Table 15.5) and delocalization is less effective in that molecule.

The addition of the doubly ionic structures to the MCVB wave function produces an energy only 0.15 eV above the full calculation and, therefore, has produced just about all the necessary delocalization.

### 15.4.2 The MOCI treatment

In this case the wave function consists of the Hartree–Fock function with added configurations involving "excitations" of electrons from the occupied to the virtual orbitals. With ten electrons we could have excitations as high as ten-fold, but we do not explicitly work out those between four-fold and the full calculation, which is, of course, the same as the full one from the MCVB. The results are shown in Table 15.13. The first thing we notice is the correct result that single excitations do not contribute to the CI energy.[6] Perhaps the next most noteworthy aspect is that the fifth through tenth excitations contribute very little to the energy lowering. Indeed, the double excitations contribute the biggest part by themselves.

The delocalization is, of course, not a problem for MOCI calculations, but the electron correlation is. The numbers show that the double excitations produce a considerable portion of the correlation energy possible with this basis, while including excitations up through quadruple produces essentially all.

---

[6] This is a consequence of Brillouin's theorem.

### *15.4.3 Conclusions*

In the introduction we pointed out that calculations in chemistry and physics frequently start from an "ideal" model and proceed to improvements. This procedure is clear in the following two cases.

**MCVB** The principal function is completely open-shell, in that it involves no electron paired in a single orbital. As ionic functions are added to the wave function, these, in many but not all cases, involve electrons paired in a single orbital and begin to contribute a closed-shell nature to the description of the system. (These ionic structures also cause delocalization, as we have seen.)

**MOCI** The principal function here is completely closed-shell and the added configurations serve to decrease this characteristic. Since the electrons become correlated under these circumstances, the delocalization is necessarily reduced.

How these two characteristics balance out depends upon the system. Nevertheless, since the MCVB "ideal" is the separated atom state, it gives a description of molecule formation that pictures molecules with more-or-less intact atoms in them.

# 16

# Interaction of molecular fragments

In previous chapters we have repeatedly emphasized that the principal difficulty in calculating the dissociation energy of a bond is the correct treatment of the change in electron correlation as the bond distance changes. This observation also applies to reactions where bonds are both formed and broken. In many important cases, however, the particular atom–atom distances that change significantly during a reaction are relatively few in number, and a method for accurately treating the correlation in only those "bonds" would have a clear advantage in efficiency. The MCVB method provides a method for targeting certain bonds to treat the correlation in them as well as possible. We call this procedure *targeted correlation*, and in this chapter we give examples using it. The SCVB method could also be used in this context.

In our previous work we have used SCF solutions of the atoms as the ingredients of the $n$-electron VB basis functions. With targeted correlation we go one step up and use SCF solutions of molecular fragments as the ingredients. As the name implies, this must be tailored to the specific example and must be done with a careful eye to the basic chemistry and physics of the situation at hand.

## 16.1 Methylene, ethylene, and cyclopropane

In this section we consider some molecules that can be viewed as consisting of methylene radicals in some combination. Earlier publications[39, 66] have covered some of the aspects of the subjects covered here. These earlier studies used an STO3G minimal basis, and provide information to make comparisons of results with the 6-31G* results that are presented here. We will describe the minimal basis results more completely in a later section. Here, however, we make one comment concerning the way one must handle these different bases. When using minimal bases with targeted correlation qualitatively reasonable results are obtained, but this is, in part, due to the less satisfactory representation of the fragments. When we use

214

a larger basis, the fragment is better described, but the orbitals are not so well conditioned to the molecules that one wishes to construct from them. Therefore, when we use a 6-31G* basis it is necessary to allow the open orbitals to breathe as distances change, as suggested by Hiberty[44]. We will discuss the methylene biradical first.

### 16.1.1 The methylene biradical

The structure of $CH_2$ was discussed by a completely MCVB treatment in Chapter 15. Here we look at it from an ROHF point of view. The structure of $CH_2$ was uncertain for a number of years, but it is now known that the ground state is triplet with a bent geometry in $C_{2v}$ symmetry. Conventions dictate that $CH_2$ be oriented with the $C_2$- and $z$-axes coincident and the molecule in the $y$–$z$ plane. Consequently the ground state is $^3B_1$, and the MO configuration is

$$1a_1^2 2a_1^2 1b_2^2 3a_1 1b_1.$$

The first excited state is the singlet configuration

$$1a_1^2 2a_1^2 1b_2^2 3a_1^2,$$

which has $^1A_1$ symmetry. The SCF energies for these are rather too far apart since there is more electron correlation in the singlet coupling. We shall be able to interpret our results for ethylene and cyclopropane in terms of these states of the methylene biradical.

### 16.1.2 Ethylene

Our treatments of ethylene are all carried out with two methylene fragments that have the $1a_1^2 2a_1^2 1b_2^2$ parts of both of their configurations doubly occupied in all VB structures used. The 12 electrons involved can be placed in the core as described in Chapter 9, which means that there are only four electrons, those for the C—C $\sigma$ and $\pi$ bonds, that are in the MCVB treatment. For simplicity we shall rename the other two methylene orbitals $\sigma_i$ and $\pi_i$, where $i = 1, 2$ for the two ends of the molecule. The Weyl dimension formula tells us that there are 20 linearly independent tableaux from four electrons distributed in four orbitals. When we use $D_{2h}$ symmetry, however, only 12 of them are involved in eight $^1A_1$ functions.

As indicated above, the $\sigma_i$ and $\pi_i$ orbitals are not the "raw" orbitals coming out of the ROHF treatment of methylene, but linear combinations of the occupied and selected virtual orbitals of that treatment, which provides the breathing adjustment. Specifically, we use

$$\sigma_i = c_1 3a_1 + c_2 4a_1 + c_3 5a_1 + c_4 6a_1 + c_5 7a_1, \tag{16.1}$$
$$\pi_i = d_1 1b_1 + d_2 2b_1 + d_3 3b_1, \tag{16.2}$$

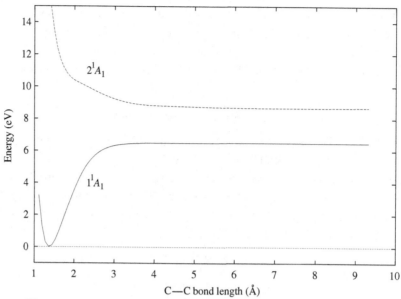

Figure 16.1. Dissociation curve for the double bond in $CH_2{=}CH_2$.

and optimize the six independent (each orbital is normalized) parameters at each C—C distance. This includes all of the $b_1$ virtual orbitals but omits the highest three $a_1$ virtual orbitals. These latter are mostly involved with the $1s$ function of the basis and will not influence bonding significantly. Figure 16.1 shows the ground and first excited singlet states of ethylene as a function of the C—C distance. The molecule is held in a plane and possesses $D_{2h}$ symmetry at all distances. The H—C—H angle, as determined from SCF minimizations, changes by about a degree in this transformation, but this nicety was not included, the angle being held at the ethylene value for all distances.

At $R_{CC} = \infty$, the ground state wave function is particularly simple in terms of standard tableaux functions,

$$\Psi_0(R = \infty) = \begin{bmatrix} \sigma_1 & \sigma_2 \\ \pi_1 & \pi_2 \end{bmatrix},\qquad(16.3)$$

where we assume the tableau symbol includes its normalization constant. This is easily interpreted as two triplet systems coupled to singlet. In terms of HLSP functions the results are not so simple. We have

$$\Psi_0(R = \infty) = 0.577\,35 \begin{bmatrix} \sigma_1 & \sigma_2 \\ \pi_1 & \pi_2 \end{bmatrix}_R - 0.577\,35 \begin{bmatrix} \sigma_2 & \pi_1 \\ \sigma_1 & \pi_2 \end{bmatrix}_R,\qquad(16.4)$$

since these do not represent the triplet states so easily. In neither case are there any ionic terms, of course.

Table 16.1. *The principal terms in the ground state wave function for $R_{CC}$ at the energy minimum. The two sorts of tableaux are given.*

| | | 1 | 2 | 3 | 4 |
|---|---|---|---|---|---|
| Standard tableaux functions | Num. | 1 | 2 | 2 | 1 |
| | Tab. | $\begin{bmatrix} \sigma_1 & \sigma_2 \\ \pi_1 & \pi_2 \end{bmatrix}$ | $\begin{bmatrix} \sigma_2 & \sigma_2 \\ \pi_1 & \pi_2 \end{bmatrix}$ | $\begin{bmatrix} \pi_2 & \pi_2 \\ \sigma_1 & \sigma_2 \end{bmatrix}$ | $\begin{bmatrix} \sigma_1 & \pi_1 \\ \sigma_2 & \pi_2 \end{bmatrix}$ |
| | $C_i(min)$ | 0.536 47 | 0.148 46 | $-0.128\,37$ | $-0.124\,77$ |
| HLSP functions | Num. | 1 | 2 | 2 | 2 |
| | Tab. | $\begin{bmatrix} \sigma_1 & \sigma_2 \\ \pi_1 & \pi_2 \end{bmatrix}_R$ | $\begin{bmatrix} \sigma_2 & \sigma_2 \\ \pi_1 & \pi_2 \end{bmatrix}_R$ | $\begin{bmatrix} \pi_1 & \pi_1 \\ \sigma_1 & \sigma_2 \end{bmatrix}_R$ | $\begin{bmatrix} \sigma_1 & \sigma_1 \\ \pi_2 & \pi_2 \end{bmatrix}_R$ |
| | $C_i(min)$ | 0.473 51 | 0.148 46 | $-0.128\,37$ | $-0.121\,06$ |

The first excited state wave function is also more complicated at $R_{CC} = \infty$. In terms of standard tableaux functions it is

$$\Psi_1(R = \infty) = 0.912\,69 \begin{bmatrix} \sigma_1 & \sigma_1 \\ \sigma_2 & \sigma_2 \end{bmatrix} - 0.399\,26 \begin{bmatrix} \sigma_1 & \sigma_1 \\ \pi_2 & \pi_2 \end{bmatrix}$$

$$- 0.399\,26 \begin{bmatrix} \sigma_2 & \sigma_2 \\ \pi_1 & \pi_1 \end{bmatrix} + 0.087\,09 \begin{bmatrix} \pi_1 & \pi_1 \\ \pi_2 & \pi_2 \end{bmatrix}. \quad (16.5)$$

The first term is the combination of two $^1A_1$ methylenes, and the others provide some electron correlation in these two structures. Since Eq. (16.5) has only doubly occupied tableaux, the HLSP functions are the same.

We show the ground state wave function at $R_{min}$ in terms of standard tableaux functions and HLSP functions in Table 16.1. We see that the representation of the wave function is quite similar in the two different ways. Considering the HLSP functions first, we note that the principal term represents two electron pair bonds, one $\sigma$ and one $\pi$. The next two are ionic structures contributing to delocalization, and the fourth is a nonionic contribution to delocalization.

The standard tableaux function representation is similar. The principal term is the same as the only term at $R = \infty$, and together with the fourth term (the other standard tableau of the constellation) represents the two electron pair bonds of the double bond. The second and third terms are the same as those in the HLSP function representation and even have the same coefficients, since there is only one function of this sort.

When we make a similar analysis of the terms in the wave function for the first excited state, more ambiguous results are obtained. These are shown in Table 16.2. For both standard tableaux functions and HLSP functions the principal structure is the same as that in the ground state. Higher terms are of an opposite sign, which provides the necessary orthogonality, but the character of the wave function is not very clear

Table 16.2. *The principal terms in the first excited state wave function for $R_{CC}$ at the energy minimum. The two sorts of tableaux are given.*

| | | 1 | 2 | 3 | 4 |
|---|---|---|---|---|---|
| Standard tableaux functions | Num. | 1 | 2 | 2 | 1 |
| | Tab. | $\begin{bmatrix} \sigma_1 & \sigma_2 \\ \pi_1 & \pi_2 \end{bmatrix}$ | $\begin{bmatrix} \pi_2 & \pi_2 \\ \sigma_1 & \sigma_2 \end{bmatrix}$ | $\begin{bmatrix} \sigma_1 & \sigma_1 \\ \pi_2 & \pi_2 \end{bmatrix}$ | $\begin{bmatrix} \sigma_1 & \sigma_1 \\ \pi_1 & \pi_2 \end{bmatrix}$ |
| | $C_i^1$ | 0.711 57 | 0.478 12 | 0.337 33 | 0.216 83 |
| HLSP functions | Num. | 1 | 2 | 2 | 2 |
| | Tab. | $\begin{bmatrix} \sigma_1 & \sigma_2 \\ \pi_1 & \pi_2 \end{bmatrix}_R$ | $\begin{bmatrix} \pi_1 & \pi_1 \\ \sigma_1 & \sigma_2 \end{bmatrix}_R$ | $\begin{bmatrix} \sigma_1 & \sigma_1 \\ \pi_2 & \pi_2 \end{bmatrix}_R$ | $\begin{bmatrix} \sigma_1 & \sigma_1 \\ \pi_1 & \pi_2 \end{bmatrix}_R$ |
| | $C_i^1$ | 0.578 01 | 0.478 12 | 0.337 33 | 0.216 83 |

from these terms. We give this example, because we will contrast it with an excited state wave function of an unambiguous sort when we discuss cyclopropane. There is one more point that should be discussed before we go on to cyclopropane, however.

The $1a_1^2 2a_1^2 1b_2^2$ orbitals on one of the two $CH_2$ fragments are not orthogonal to those on the other $CH_2$, but this does not cause the core valence separation any problems. It does, however, represent a repulsion: that which is normally expected between closed-shell systems. In this case the overlap and the repulsion are small. During the calculations of the core matrix elements a measure of the overlap is computed. This number is exactly 1.0 if there is zero overlap between the fragment core orbitals. As more overlap appears the number rises and can be as high as 100. For ethylene we never press the $CH_2$ fragments close enough together to reach numbers higher than about 1.02. This is quite small and is the reason we know that the core repulsion is small in this system.

### 16.1.3 Cyclopropane with a 6-31G* basis

We examine the two lowest singlet states of cyclopropane as one of the $CH_2$ groups is pulled away from the other two. Figure 16.2 shows the basic arrangement of the molecule with the three C atoms in the $x$–$y$ plane. The C atom on the right is on the $y$-axis, and $R_1$ is its distance to the midpoint of the other two Cs. $R_2$ is the distance between the two Cs that will become part of ethylene and $\phi$ is the angle out of planarity. We have labeled the C atoms 1, 2, and 3 to identify the three different methylenes for designating orbitals.

The wave function is an extension of the one we used for the dissociation of ethylene. We now have 18 electrons in nine core orbitals, and six electrons in the three $\sigma$ and three $\pi$ orbitals that will make up the C—C bonds. As before, the valence orbitals are allowed to breathe (see Eqs. (16.1) and (16.2) for the linear combinations) as the system changes. According to the Weyl dimension formula

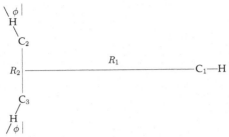

Figure 16.2. The geometric arrangement of the atoms of cyclopropane during the dissociation to ethylene and methylene. The system is maintained in $C_{2v}$ geometry as $R_1$, $R_2$, and $\phi$ change.

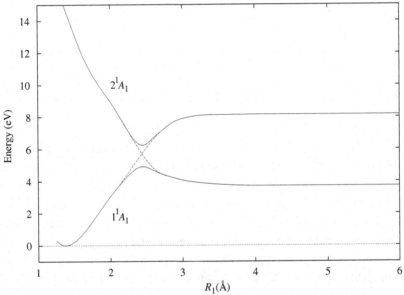

Figure 16.3. The two lowest $^1A_1$ states during the dissociation of cyclopropane along the $C_2H_4$ relaxed path. The dashed lines, indicating the diabatic energies, were not computed but have been added merely to guide the eye.

there are 175 different structures possible, but in this case there are only 173 of them involved in 92 $^1A_1$ ($C_{2v}$) symmetry functions. The core overlap criterion never becomes larger than 1.021 for these calculations.

In Fig. 16.3 we show the two lowest $^1A_1$ states as a function of $R_1$ for optimum values of the $R_2$ and $\phi$ parameters, and in Fig. 16.4 we show the path by giving $R_2$ and $\phi$ as functions of $R_1$. Qualitatively, the two energy curves have the classic appearance of an *avoided crossing* between two diabatic states. The dashed lines in Fig. 16.3 are not computed, but have been added to guide the eye.

Although we have been speaking of our process as the dissociation of cyclopropane, it is simpler to discuss it as if proceeding from the other direction.

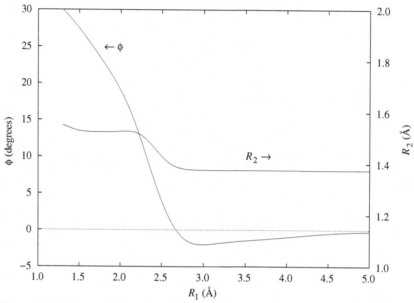

Figure 16.4. The variation of $R_2$ and $\phi$ during the dissociation of cyclopropane.

Considering the dissociated system, we see that the geometry has relaxed to that of an ethylene molecule with a methylene radical at some distance. Our treatment allows methylene to have only two electrons in open shells so it must be in either a singlet or a triplet state. Since the overall system is in a singlet state, the ethylene portion must also be either singlet or triplet, respectively, to match. At the same time ethylene is singlet in its ground state and triplet in its first excited state with a fairly large excitation energy, while methylene is triplet in its ground state and singlet in its first excited state with a relatively small excitation energy. These facts tell us that, at long distances, the lower $^1A_1$ state is a combination of ground state ethylene and singlet methylene. The Rumer tableau that corresponds to this case is

$$
\begin{bmatrix}
\sigma_1 & \sigma_1 \\
\sigma_2 & \sigma_3 \\
\pi_2 & \pi_3
\end{bmatrix}_R ,
$$

where the subscripts identify the particular methylene fragment as given in Fig. 16.2. There is, of course, another Rumer structure corresponding to this orbital set. Table 16.3 gives the coefficients in the wave function.

At long distances the first excited $^1A_1$ state, on the other hand, is a combination of triplet ethylene and ground state (triplet) methylene. The important tableaux in the wave function are shown in Table 16.4, and in this case the principal tableau of

Table 16.3. *The leading tableaux for the ground state wave function*
*of $C_2H_4 + CH_2$ at infinite separation.*

| | | 1 | 2 | 3 | 4 |
|---|---|---|---|---|---|
| Standard tableaux functions | Num. | 1 | 1 | 2 | 2 |
| | Tab. | $\begin{bmatrix} \sigma_1 & \sigma_1 \\ \sigma_2 & \sigma_3 \\ \pi_2 & \pi_3 \end{bmatrix}$ | $\begin{bmatrix} \pi_1 & \pi_1 \\ \sigma_2 & \sigma_3 \\ \pi_2 & \pi_3 \end{bmatrix}$ | $\begin{bmatrix} \sigma_1 & \sigma_1 \\ \sigma_2 & \sigma_2 \\ \pi_2 & \pi_3 \end{bmatrix}$ | $\begin{bmatrix} \sigma_1 & \sigma_1 \\ \sigma_2 & \pi_2 \\ \sigma_3 & \pi_3 \end{bmatrix}$ |
| | $C_i(\infty)$ | 0.529 89 | $-0.156\,11$ | 0.146 58 | $-0.128\,41$ |
| HLSP functions | Num. | 1 | 2 | 1 | 2 |
| | Tab. | $\begin{bmatrix} \sigma_1 & \sigma_1 \\ \sigma_2 & \sigma_3 \\ \pi_2 & \pi_3 \end{bmatrix}_R$ | $\begin{bmatrix} \sigma_1 & \sigma_1 \\ \sigma_2 & \sigma_2 \\ \pi_2 & \pi_3 \end{bmatrix}_R$ | $\begin{bmatrix} \pi_1 & \pi_1 \\ \sigma_2 & \sigma_3 \\ \pi_2 & \pi_3 \end{bmatrix}_R$ | $\begin{bmatrix} \sigma_1 & \sigma_1 \\ \pi_2 & \pi_2 \\ \sigma_2 & \sigma_3 \end{bmatrix}_R$ |
| | $C_i(\infty)$ | 0.463 18 | 0.146 58 | $-0.136\,47$ | $-0.120\,81$ |

Table 16.4. *The leading tableaux for the first excited state wave function*
*of $C_2H_4 + CH_2$ at infinite separation.*

| | | 1 | 2 | 3 | 4 |
|---|---|---|---|---|---|
| Standard tableaux functions | Num. | 1 | 1 | 2 | 2 |
| | Tab. | $\begin{bmatrix} \sigma_1 & \sigma_2 \\ \sigma_3 & \pi_2 \\ \pi_1 & \pi_3 \end{bmatrix}$ | $\begin{bmatrix} \sigma_1 & \sigma_3 \\ \sigma_2 & \pi_2 \\ \pi_1 & \pi_3 \end{bmatrix}$ | $\begin{bmatrix} \sigma_3 & \sigma_3 \\ \sigma_1 & \pi_2 \\ \pi_1 & \pi_3 \end{bmatrix}$ | $\begin{bmatrix} \sigma_1 & \sigma_3 \\ \sigma_2 & \pi_1 \\ \pi_2 & \pi_3 \end{bmatrix}$ |
| | $C_i(\infty)$ | 0.414 52 | 0.325 77 | $-0.206\,49$ | 0.096 43 |
| HLSP functions | Num. | 1 | 2 | 1 | 2 |
| | Tab. | $\begin{bmatrix} \sigma_2 & \sigma_3 \\ \pi_1 & \pi_2 \\ \sigma_1 & \pi_3 \end{bmatrix}_R$ | $\begin{bmatrix} \sigma_2 & \sigma_3 \\ \sigma_1 & \pi_1 \\ \pi_2 & \pi_3 \end{bmatrix}_R$ | $\begin{bmatrix} \sigma_3 & \sigma_3 \\ \pi_1 & \pi_2 \\ \sigma_1 & \pi_3 \end{bmatrix}_R$ | $\begin{bmatrix} \sigma_3 & \sigma_3 \\ \sigma_1 & \pi_1 \\ \pi_2 & \pi_3 \end{bmatrix}_R$ |
| | $C_i(\infty)$ | 0.801 48 | 0.383 16 | 0.245 93 | 0.133 59 |

the state is more easily written[1] in terms of standard tableaux functions as

$$\begin{bmatrix} \sigma_1 & \sigma_2 \\ \sigma_3 & \pi_2 \\ \pi_1 & \pi_3 \end{bmatrix} = - \begin{bmatrix} \sigma_2 & \sigma_3 \\ \pi_2 & \sigma_1 \\ \pi_3 & \pi_1 \end{bmatrix},$$

where there are, of course, four more standard tableaux that could be written. We may actually use the tableau shapes to see that this is the correct interpretation of

---

[1] In this book the entries in tables like Table 16.4 are generated semiautomatically from computer printout. Although this almost completely eliminates the dangers of misprints the computer programs do not always arrange the tableaux in the most convenient way for the discussion. In this case we use the properties of standard tableaux functions to make the transformation used.

Table 16.5. *The leading tableaux for the ground state wave function of $C_3H_6$ at the equilibrium geometry.*

|  |  | 1 | 2 | 3 | 4 |
|---|---|---|---|---|---|
| Standard tableaux functions | Num. | 1 | 1 | 1 | 1 |
|  | Tab. | $\begin{bmatrix} \sigma_1 & \sigma_2 \\ \sigma_3 & \pi_2 \\ \pi_1 & \pi_3 \end{bmatrix}$ | $\begin{bmatrix} \sigma_1 & \sigma_3 \\ \sigma_2 & \pi_1 \\ \pi_2 & \pi_3 \end{bmatrix}$ | $\begin{bmatrix} \sigma_1 & \sigma_2 \\ \sigma_3 & \pi_1 \\ \pi_2 & \pi_3 \end{bmatrix}$ | $\begin{bmatrix} \pi_1 & \pi_1 \\ \sigma_2 & \sigma_3 \\ \pi_2 & \pi_3 \end{bmatrix}$ |
|  | $C_i(min)$ | 0.134 41 | 0.130 92 | $-0.123\ 14$ | 0.102 56 |
| HLSP function | Num. | 1 | 1 | 1 | 2 |
|  | Tab. | $\begin{bmatrix} \sigma_2 & \sigma_3 \\ \pi_1 & \pi_2 \\ \sigma_1 & \pi_3 \end{bmatrix}_R$ | $\begin{bmatrix} \sigma_1 & \sigma_2 \\ \sigma_3 & \pi_1 \\ \pi_2 & \pi_3 \end{bmatrix}_R$ | $\begin{bmatrix} \pi_1 & \pi_1 \\ \sigma_2 & \sigma_3 \\ \pi_2 & \pi_3 \end{bmatrix}_R$ | $\begin{bmatrix} \pi_3 & \pi_3 \\ \sigma_1 & \sigma_2 \\ \pi_1 & \pi_2 \end{bmatrix}_R$ |
|  | $C_i(min)$ | 0.132 70 | $-0.125\ 19$ | 0.086 67 | 0.085 38 |

the states of the separate pieces of our system. The last tableau given above can also be written symbolically as

$$\begin{bmatrix} \sigma_2 & \sigma_3 \\ \pi_2 \\ \pi_3 \end{bmatrix} + \begin{bmatrix} \sigma_1 \\ \pi_1 \end{bmatrix},$$

where the two triplet tableaux can fit together to form the earlier singlet shape. We have not emphasized this sort of combining of tableaux in our earlier work, but it is particularly useful for systems in asymptotic regions. There are more structures for this state and set of orbitals than for the lower one, because this set must also represent a still higher coupling of ethylene and methylene, both in a $^1B_1$ state. We do not show the energy curve for the state that goes asymptotically to that coupling.

The wave functions for the ground and first excited $^1A_1$ states for the cyclopropane equilibrium geometry are shown in Tables 16.5 and 16.6. In the case of the ground state either the principal standard tableaux function or HLSP function can be transformed as follows:

$$\begin{bmatrix} \sigma_1 & \sigma_2 \\ \sigma_3 & \pi_2 \\ \pi_1 & \pi_3 \end{bmatrix} = -\begin{bmatrix} \sigma_2 & \sigma_3 \\ \pi_2 & \sigma_1 \\ \pi_3 & \pi_1 \end{bmatrix},$$

$$\begin{bmatrix} \sigma_2 & \sigma_3 \\ \pi_1 & \pi_2 \\ \sigma_1 & \pi_3 \end{bmatrix}_R = \begin{bmatrix} \sigma_2 & \sigma_3 \\ \pi_2 & \sigma_1 \\ \pi_3 & \pi_1 \end{bmatrix}_R.$$

Table 16.6. *The leading tableaux for the first excited state wave function of $C_3H_6$ at the equilibrium geometry.*

|  |  | 1 | 2 | 3 | 4 |
|---|---|---|---|---|---|
| Standard tableaux functions | Num. | 1 | 1 | 1 | 1 |
|  | Tab. | $\begin{bmatrix} \pi_1 & \pi_1 \\ \sigma_2 & \sigma_3 \\ \pi_2 & \pi_3 \end{bmatrix}$ | $\begin{bmatrix} \sigma_1 & \sigma_3 \\ \sigma_2 & \pi_2 \\ \pi_1 & \pi_3 \end{bmatrix}$ | $\begin{bmatrix} \sigma_1 & \sigma_3 \\ \sigma_2 & \pi_1 \\ \pi_2 & \pi_3 \end{bmatrix}$ | $\begin{bmatrix} \sigma_1 & \sigma_2 \\ \sigma_3 & \pi_1 \\ \pi_2 & \pi_3 \end{bmatrix}$ |
|  | $C_i(min)$ | 0.205 06 | $-0.182\,87$ | 0.176 27 | $-0.165\,80$ |
| HLSP functions | Num. | 1 | 1 | 1 | 1 |
|  | Tab. | $\begin{bmatrix} \sigma_2 & \sigma_3 \\ \pi_1 & \pi_2 \\ \sigma_1 & \pi_3 \end{bmatrix}_R$ | $\begin{bmatrix} \pi_1 & \pi_1 \\ \sigma_2 & \sigma_3 \\ \pi_2 & \pi_3 \end{bmatrix}_R$ | $\begin{bmatrix} \sigma_1 & \sigma_2 \\ \sigma_3 & \pi_1 \\ \pi_2 & \pi_3 \end{bmatrix}_R$ | $\begin{bmatrix} \sigma_2 & \sigma_3 \\ \sigma_1 & \pi_1 \\ \pi_2 & \pi_3 \end{bmatrix}_R$ |
|  | $C_i(min)$ | 0.182 33 | $-0.173\,96$ | 0.168 55 | 0.167 47 |

The first of these is the same as the transformation above for the first excited asymptotic wave function. Thus, at the minimum geometry we see that the leading term in the wave function is the same as that for the first excited state at $\infty$, and there has been a cross-over in the character of the wave function for the two geometries.

The leading coefficients are rather small for these functions. This is in part because the orbital set in terms of which we have expressed the functions is not the most felicitous. We have used $\sigma_i$ and $\pi_i$ relating to the local geometry of each methylene. Alternatively, we can form hybrid orbitals

$$h_{i1} = N(\sigma_i + \pi_i),$$
$$h_{i2} = N(\sigma_i - \pi_i),$$

which, in each case, are directed towards a neighboring methylene. The signs of the orbitals are such that these combinations yield

$$h_{12} \leftrightarrow h_{21},$$
$$h_{22} \leftrightarrow h_{31},$$
$$h_{32} \leftrightarrow h_{11},$$

as the pairs that overlap most strongly. Table 16.7 shows the HLSP function tableaux for the ground state in terms of these hybrids. This representation gives little clue as to the asymptotic state this might be connected with, but does show a rather conventional picture of cyclopropane as having three electron pair bonds holding the ring together. There is also the expected mix of covalent and ionic functions.

When we get to the first excited state at the geometry of the energy minimum (Table 16.6), it is seen that most important tableaux in the wave function in terms

Table 16.7. *The leading HLSP functions for the ground state wave function of $C_3H_6$ at the equilibrium geometry when hybrid orbitals are used.*

|  | 1 | 2 | 3 | 4 |
|---|---|---|---|---|
| Num. | 1 | 6 | 6 | 2 |
| Tab. | $\begin{bmatrix} h_{32} & h_{11} \\ h_{12} & h_{21} \\ h_{22} & h_{31} \end{bmatrix}_R$ | $\begin{bmatrix} h_{11} & h_{11} \\ h_{12} & h_{21} \\ h_{22} & h_{31} \end{bmatrix}_R$ | $\begin{bmatrix} h_{11} & h_{11} \\ h_{31} & h_{31} \\ h_{12} & h_{21} \end{bmatrix}_R$ | $\begin{bmatrix} h_{11} & h_{11} \\ h_{21} & h_{21} \\ h_{31} & h_{31} \end{bmatrix}_R$ |
| $C_i(min)$ | 0.354 55 | 0.098 99 | 0.052 40 | 0.046 07 |

of $\sigma_i$ and $\pi_i$ orbitals is not very easily interpreted, although the leading term, in the case of HLSP functions, is the same as that for the ground state. We do not give them here, but the first excited state in terms of the hybrid orbitals is likewise poorly illuminating. We may look at the problem in another way.

As cyclopropane dissociates, we see that the geometry changes happen rather rapidly over a fairly narrow range as the character of the energy states changes in the neighborhood of $R_1 = 2.4\,\text{Å}$. (See Fig. 16.4.) At asymptotic geometries we saw that the characters of the wave functions for the first two states are clearcut. As the one methylene moves, the two pieces in the first excited state, consisting of two triplet fragments, attract one another more strongly and the potential energy curve falls, see Fig. 16.3. The ground state, consisting of two singlet fragments appears repulsive. These two sorts of states would cross if they did not interact. They, in fact, do interact: there is an avoided crossing, and a barrier appears on the lower curve. This interaction region is fairly narrow, and, inside the cross-over, the lower curve continues downward representing the bonding that holds $C_3H_6$ together. Thus, this targeted correlation treatment predicts that there is a 1.244 eV barrier to the insertion of singlet methylene into ethylene to form cyclopropane. We do not show it here[39], but triplet methylene and singlet ethylene repel each other strongly at all distances, and thus should not react unless there should be a spin cross-over to a singlet state. This occurrence of a barrier due to an avoided crossing has been invoked many times to explain and rationalize reaction pathways [67, 69].

### 16.1.4 Cyclopropane with an STO-3G basis

Some years ago a short description of a more restricted version of the problem in the last section was published[39]. Using an STO-3G basis, the earlier calculation examined the two lowest $^1A_1$ energies as a singlet methylene approached an ethylene molecule. In this case, however, the ethylene was not allowed to relax in its geometry. The curves are shown in Fig. 16.5. The important point is that we see

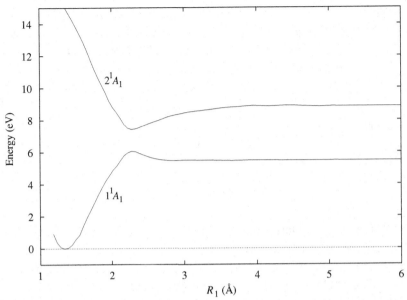

Figure 16.5. The two lowest $^1A_1$ states showing the attack of singlet methylene on a rigid ethylene. These energies were obtained using an STO-3G basis, with which we obtain a barrier of about 0.8 eV.

the same qualitative behavior in this much more approximate calculation as that shown in Fig. 16.3, where the results using a larger basis and fuller optimization is presented.

## 16.2 Formaldehyde, $H_2CO$

When formaldehyde is subjected to suitable optical excitation it dissociates into $H_2$ and CO. The process is thought to involve an excitation to the first excited singlet state followed by internal conversion to a highly excited vibrational state of the ground singlet state that dissociates according to the equation

$$H_2CO \xrightarrow{h\nu} H_2CO^* \rightarrow H_2 + CO.$$

Conventional counting says that $H_2CO$ has four bonds in it, and the final product has the same number arranged differently. Our goal is to follow the bonding arrangement from the initial geometry to the final. This is said to occur on the $S_0$ (ground state singlet) energy surface, which in full generality depends upon six geometric parameters. Restricting the surface to planar geometries reduces this number to five, and keeping the C—O distance fixed reduces it to four. We will examine different portions of the $S_0$ surface for different numbers of geometric coordinates.

Some years ago Vance and the present author[68] made a study of this surface with the targeted correlation technique using a Dunning double-*zeta* basis[70] that,

Table 16.8. *Orbitals used and statistics of MCVB calculations on the $S_0$ energy surface of formaldehyde. For the 6-31G\* basis the full set of configurations from the unprimed orbitals was used and single excitations into the primed set were included.*

|  | Orbitals | Total | $C_{2v}$ | $C_s$ |
|---|---|---|---|---|
| STO3G basis | H:$1s_a$, $1s_b$ |  |  |  |
|  | CO:$3\sigma$, $3\sigma$, $5\sigma$, $6\sigma$, $1\pi$, $2\pi$ | 1120 | 565 | 1120 |
| 6-31G\* basis | H:$1s_a$, $1s_b$, $2s'_a$, $2s'_b$ |  |  |  |
|  | CO:$5\sigma$, $1\pi$, $5\sigma'$, $2\pi$, $2\pi'$ | 131 | 70 | 131 |

except for the lack of polarization functions, is similar to a 6-31G\* basis. To keep consistency with the remainder of this book we redo some of the calculations from the earlier study with the latter basis, but will mix in some of the earlier results, which are essentially the same, with the current ones.

We show the results of calculations at the STO3G and 6-31G\* levels of the AO basis. Table 16.8 shows the orbitals used and the number of functions produced for each case. These statistics apply to each of the calculations we give.

The important difference between the STO3G and 6-31G\* bases is the arrangement of orbitals on the CO fragment. In its ground state CO has an orbital configuration of

$$\text{Core: } 3\sigma^2 4\sigma^2 5\sigma^2 1\pi^4.$$

The $5\sigma$ function is best described as a nonbonding orbital located principally on the C atom. In Table 16.8 the $2\pi$ orbital is the virtual orbital from the ground state RHF treatment. The primed orbitals on H are the same as we have used before, but those on CO are based upon an ROHF $n \rightarrow \pi^*$ calculation of the first triplet state. The "raw" $5\sigma$, $5\sigma'$, $2\pi$, and $2\pi'$ taken directly from the calculations will not work, however. Their overlaps are much too large for an $S$ matrix of any size ($>2$ or 3) to be considered nonsingular by standard 16-place accuracy calculations. Therefore, for each high-overlap pair the sum and difference were formed. These are orthogonal, and do not cause any problems.

### 16.2.1 The least motion path

We first comment on the so-called least motion path (LMP), in which the two H atoms move away from the CO atoms, maintaining a $C_{2v}$ symmetry, as shown in Fig. 16.6. Earlier calculations of all sorts indicate that this path does not cross

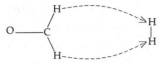

Figure 16.6. A representation of the LMP for the dissociation of H$_2$CO.

Figure 16.7. The true saddle point for the dissociation of H$_2$CO. This figure is drawn to scale as accurately as possible.

the lowest saddle point for the reaction. In fact, there is no real saddle point in geometries constrained to be $C_{2v}$. Earlier workers have, however, imposed further constraints and produced a pseudo saddle point of this sort. This is done because it illustrates a typical four-electron rearrangement similar to the process discussed in Chapter 14 for four H atoms. This is classified by Woodward and Hoffman[58] as a "forbidden" process, which means, of course, that the energy required for it is relatively high compared to the energy for other geometries that may break the symmetry giving the orbital crossing. In any event the forces on the nuclei along restricted paths such as this tend to lead to separation of all three parts of the molecule rather than the formation of CO and H$_2$.

### 16.2.2 The true saddle point

Calculations using both MCVB and MOCI wave functions predict a very different geometry at the saddle point for the H$_2$CO dissociation. The molecule is still planar, but otherwise has no elements of symmetry. We do not describe calculations here that search out the saddle point, but we do show the nature of the wave function there, which will make clear why it has the relatively peculiar geometry shown in Fig. 16.7. This position is such that the tendency of the molecule is to form a H$_2$ molecule. Depending upon the method of calculation the barrier height is estimated to be 4.05–4.06 eV, approximately the energy of one H—H bond. Theoretically[2] the exothermicity of the process is very close to 0.0 so the parts separate with at least the activation energy.

---

[2] The process we are discussing is a so-called *isodesmic* reaction. This means that the number of bonds is constant. It has been argued that calculations of this sort of process using changes in SCF energies are useful because the correlation energies tend to cancel when taking the difference. See Ref. [71].

Table 16.9. *The leading Rumer tableaux for the asymptotic state of formaldehyde,*
$H_2 + CO$. *In this case the standard tableaux functions*
*are the same for these terms.*

|  | 1 | 2 | 3 | 4 |
|---|---|---|---|---|
| Num. | 1 | 2 | 2 | 2 |
| Tab. | $\begin{bmatrix} 5\sigma & 5\sigma \\ 1\pi_x & 1\pi_x \\ 1\pi_y & 1\pi_y \\ 1s_a & 1s_b \end{bmatrix}_R$ | $\begin{bmatrix} 1s_a & 1s_a \\ 5\sigma & 5\sigma \\ 1\pi_x & 1\pi_x \\ 1\pi_y & 1\pi_y \end{bmatrix}_R$ | $\begin{bmatrix} 5\sigma & 5\sigma \\ 1\pi_x & 1\pi_x \\ 1\pi_y & 1\pi_y \\ 1s_a & 1s_b' \end{bmatrix}_R$ | $\begin{bmatrix} 5\sigma & 5\sigma \\ 1\pi_x & 1\pi_x \\ 2\pi_y & 2\pi_y \\ 1s_a & 1s_b \end{bmatrix}_R$ |
| $C_i(inf)$ | 0.709 68 | 0.168 56 | $-0.134\,08$ | $-0.098\,04$ |

Table 16.10. *The leading Rumer tableaux in the wave function for the saddle*
*point state of formaldehyde dissociation.*

|  | 1 | 2 | 3 | 4 |
|---|---|---|---|---|
| Num. | 1 | 1 | 1 | 1 |
| Tab. | $\begin{bmatrix} 1\pi_x & 1\pi_x \\ 1\pi_y & 1\pi_y \\ 1s_b & 5\sigma' \\ 1s_a & 2\pi_y' \end{bmatrix}_R$ | $\begin{bmatrix} 1\pi_x & 1\pi_x \\ 1\pi_y & 1\pi_y \\ 1s_a & 1s_b \\ 5\sigma' & 2\pi_y' \end{bmatrix}_R$ | $\begin{bmatrix} 1\pi_x & 1\pi_x \\ 1s_b & 5\sigma' \\ 1s_a & 1\pi_y \\ 2\pi_y & 2\pi_y' \end{bmatrix}_R$ | $\begin{bmatrix} 1\pi_x & 1\pi_x \\ 2\pi_y' & 2\pi_y' \\ 1s_b & 5\sigma' \\ 1s_a & 1\pi_y \end{bmatrix}_R$ |
| $C_i(sad)$ | 1.306 47 | 0.461 32 | 0.403 68 | 0.209 90 |

### 16.2.3 Wave functions during separation

The wave functions change character, of course, during the dissociation process. The asymptotic region is the simplest and we start with that. Table 16.9 shows the most important Rumer tableaux when CO and $H_2$ are well separated from one another.

1. The leading term is clearly the closed shell $^1A_1$ state of CO in combination with the HLSP function for $H_2$.
2. These two terms give the closed-shell CO with the ionic term of $H_2$.
3. These two terms give the closed-shell CO with a breathing term for the H1$s$ orbitals.
4. The last two terms shown give the electron correlation in the $\pi$ shell of CO and the leading HLSP function term of $H_2$.

The wave function for this geometry is very simple to interpret.

In Table 16.10 we show the principal Rumer tableaux for the wave function at the saddle point.

Table 16.11. *The leading terms in the wave function for the equilibrium geometry of formaldehyde.*

|  |  | 1 | 2 | 3 | 4 |
|---|---|---|---|---|---|
| Standard tableaux functions | Num. | 1 | 2 | 2 | 2 |
|  | Tab. | $\begin{bmatrix} 1\pi_x & 1\pi_x \\ 1\pi_y & 1\pi_y \\ 1s_a & 5\sigma' \\ 1s_b & 2\pi'_y \end{bmatrix}$ | $\begin{bmatrix} 5\sigma' & 5\sigma' \\ 1\pi_x & 1\pi_x \\ 1\pi_y & 1\pi_y \\ 1s_a & 2\pi'_y \end{bmatrix}$ | $\begin{bmatrix} 1\pi_x & 1\pi_x \\ 1\pi_y & 1\pi_y \\ 2\pi'_y & 2\pi'_y \\ 1s_b & 5\sigma' \end{bmatrix}$ | $\begin{bmatrix} 1\pi_x & 1\pi_x \\ 1\pi_y & 1\pi_y \\ 1s_b & 5\sigma' \\ 2\pi'_y & 1s_a' \end{bmatrix}$ |
|  | $C_i(min)$ | 0.296 23 | 0.178 83 | 0.136 65 | 0.119 13 |
| HLSP functions | Num. | 1 | 2 | 2 | 2 |
|  | Tab. | $\begin{bmatrix} 1\pi_x & 1\pi_x \\ 1\pi_y & 1\pi_y \\ 1s_b & 5\sigma' \\ 1s_a & 2\pi'_y \end{bmatrix}_R$ | $\begin{bmatrix} 5\sigma' & 5\sigma' \\ 1\pi_x & 1\pi_x \\ 1\pi_y & 1\pi_y \\ 1s_b & 2\pi'_y \end{bmatrix}_R$ | $\begin{bmatrix} 1\pi_x & 1\pi_x \\ 1\pi_y & 1\pi_y \\ 1s_a & 5\sigma' \\ 2\pi'_y & 1sb' \end{bmatrix}_R$ | $\begin{bmatrix} 1\pi_x & 1\pi_x \\ 1\pi_y & 1\pi_y \\ 2\pi'_y & 2\pi'_y \\ 1s_b & 5\sigma' \end{bmatrix}_R$ |
|  | $C_i(min)$ | 0.310 08 | 0.178 83 | 0.159 18 | −0.136 65 |

1. The leading term represents a structure with an electron pair bond between one H and the $5\sigma'$ orbital and another between the other H and the $2\pi'_y$ orbital.
2. This term together with the first provides the two Rumer diagrams for the bonding scheme. That this has such a large coefficient indicates that electron pair bonds are not near perfect pairing.
3. This term involves correlation and polarization on the CO portion of the system with the electron pair bonds to the Hs still in place.
4. The fourth term involves a further rearrangement of the electrons on the CO portion of the system. In this case one H is now bonded to the $1\pi_y$ instead of the $2\pi'_y$ orbital. These terms provide correlation, polarization, and also give a combination of both bonding and antibonding $\pi_y$ orbitals so that this sort of bond will disappear and C—H bonds and O nonbonding orbitals will appear as the molecule forms.

Table 16.11 shows the leading terms in the wave function at the equilibrium geometry of $H_2CO$ in both standard tableaux function and HLSP function form.

1. The first standard tableaux function term is essentially triplet $H_2$ (much elongated, of course) coupled with the $^3\Pi_y$ state of CO. The HLSP function has the same interpretation.
2. The second terms are the same and are an ionic type associated with the first term. These provide delocalization.
3. The third standard tableaux function and fourth HLSP function terms are the same. These are both ionic and provide antibonding character in the y-direction to "remove" that part of the original triple bond in CO.
4. The fourth standard tableaux function term and the third HLSP function term are the same configuration but not the same function. In both cases, however, the terms involve breathing for the $1s$ orbitals in the H atoms.

In summary we see that the barrier to dissociation in $H_2CO$ can be ascribed to an avoided crossing of the same sort as we described in the dissociation of cyclopropane. The two fragments in triplet couplings bond as they approach and that state crosses the state where they are separately in singlet states. At the saddle point position the triplet fragment states still dominate to some extent, but asymptotically the two fragments are certainly in their respective singlet states.

# References

[1] L. Pauling and Jr E. B. Wilson. *Introduction to Quantum Mechanics*. McGraw–Hill Book Co., New York (1935).

[2] H. Eyring, J. Walter, and G. E. Kimball. *Quantum Chemistry*. John Wiley and Sons, Inc., New York (1944).

[3] I. N. Levine. *Quantum Chemistry (Rev.)*. Allyn and Bacon, Boston (1974).

[4] A. Messiah. *Quantum Mechanics*, Vols. I and II. North–Holland Publishing Co., Amsterdam (1966).

[5] F. A. Cotton. *Chemical Applications of Group Theory*, 2nd Ed. John Wiley and Sons, Inc., New York (1969).

[6] M. Hammermesh. *Group Theory*. Addison–Wesley, Reading, Mass (1962).

[7] D. E. Rutherford. *Substitutional Analysis*. Edinburgh University Press, reprinted by Hafner, New York (1968).

[8] W. Heitler and F. London. *Z. Physik*, **44**, 619 (1927).

[9] F. Hund. *Z. Physik*, **40**, 742; **42**, 93 (1927).

[10] R. S. Mulliken. *Phys. Rev.*, **32**, 186, 761 (1928).

[11] G. Rumer. *Göttinger Nachr.*, 377 (1932).

[12] R. Serber. *Phys. Rev.*, **45**, 461 (1934).

[13] J. H. Van Vleck and A. Sherman. *Rev. Mod. Phys.*, **7**, 167 (1935).

[14] M. Kotani, A. Amemiya, E. Ishiguro, and T. Kimura. *Table of Molecular Integrals*. Maruzen Co. Ltd., Tokyo (1955).

[15] C. A. Coulson and I. Fisher. *Phil. Mag.*, **40**, 386 (1949).

[16] H. B. G. Casimir and D. Polder. *Phys. Rev.*, **73**, 360 (1948).

[17] P. A. M. Dirac. *Proc. Roy. Soc. (London)*, **123**, 714 (1929).

[18] M. Born and J. R. Oppenheimer. *Ann. Physik*, **84**, 457 (1927).

[19] M. Born and K. Huang. *Dynamical Theory of Crystal Lattices*. Clarendon Press, Oxford (1954).

[20] R. Englman. *The Jahn Teller Effect in Molecules and Crystals*. Wiley-Interscience, New York (1972).

[21] L. Brillouin, *Jour. de Physique*, **3**, 379 (1932).

[22] J. K. L. MacDonald, *Phys. Rev.*, **43**, 830 (1933).

[23] B. H. Chirgwin and C. A. Coulson. *Proc. Roy. Soc. (London)*, **A201**, 196 (1950).

[24] G. A. Gallup and J. M. Norbeck. *Chem. Phys. Lett.*, **21**, 495 (1973).

[25] P.-O. Löwdin. *Ark. Mat. Astr. Fysk.*, **35A**, 9 (1947).

[26] C. Herring. *Rev. Mod. Phys.*, **34**, 631 (1964).

[27] J. C. Slater. *Quantum Theory of Molecules and Solids*, Vol. 1. McGraw-Hill Book Co., New York (1963).

[28] M. Abramowitz and I. A. Stegun. *Handbook of Mathematical Functions*. National Bureau of Standards, US Government Printing Office (1970).

[29] S. C. Wang. *Phys. Rev.*, **31**, 579 (1928).

[30] S. Weinbaum. *J. Chem. Phys*, **1**, 593 (1933).

[31] W. Kolos and L. Wolniewicz. *J. Chem. Phys.*, **41**, 3663 (1964).

[32] N. Rosen. *Phys. Rev.*, **38**, 2099 (1931).

[33] S. F. Boys. *Proc. Roy. Soc. (London)*, **A200**, 542 (1950).

[34] G. G. Balint-Kurti and M. Karplus. *J. Chem. Phys.*, **50**, 478 (1968).

[35] G. Wannier. *Phys. Rev.*, **52**, 191 (1937).

[36] R. Pauncz. *The Construction of Spin Eigenfunctions*. Kluwer Academic/Plenum, New York (2000).

[37] D. E. Littlewood. *The Theory of Group Characters*. Oxford University Press, London (1950).

[38] H. Weyl. *The Theory of Groups and Quantum Mechanics*. Dover Publications, Inc., London (1931).

[39] G. A. Gallup, R. L. Vance, J. R. Collins, and J. M. Norbeck. *Ad. Quantum Chem.*, **16**, 229 (1982).

[40] F. A. Matsen. *Ad. Quantum Chem.*, **1**, 60 (1964).

[41] W. A. Goddard III. *Phys. Rev.*, **157**, 81 (1967).

[42] A. C. Aitken. *Determinants and Matrices*. Interscience Publishers, New York (1956).

[43] G. A. Gallup. *Intern. J. Quantum Chem.*, **16**, 267 (1974).

[44] P. C. Hiberty. *THEOCHEM*, **398–399**, 35 (1997).

[45] C. C. J. Roothaan. *Rev. Mod. Phys.*, **32**, 179 (1960).

[46] M. W. Schmidt, K. K. Baldridge, J. A. Boatz, S. T. Elbert, M. S. Gordon, J. H. Jensen, S. Koseki, N. Matsunaga, K. A. Nguyen, S. J. Su, T. L. Windus, M. Dupuis, and J. A. Montgomery. *J. Comput. Chem.*, **14**, 1347 (1993).

[47] D. Chen and G. A. Gallup. *J. Chem. Phys.*, **93**, 8893 (1990).

[48] S. Huzinaga. *Gaussian Basis Sets for Molecular Calculations*. Elsevier Science Publishing Co., New York (1984).

[49] K. P. Huber and G. Herzberg. *Molecular Spectra and Molecular Structure. IV. Constants of Diatomic Molecules*. Van Nostrand Reinhold, New York (1979).

[50] L. Pauling. *The Nature of the Chemical Bond*. Cornell University Press, Ithica (1960).

[51] R. S. Mulliken. *J. Chem. Phys.*, **2**, 782 (1934).

[52] L. Allen. *J. Amer. Chem. Soc.*, **111**, 9003 (1989).

[53] S. Huzinaga, E. Miyoshi, and M. Sekiya. *J. Comp. Chem.*, **14**, 1440 (1993).

[54] G. Herzberg. *Molecule Spectra and Molecular Structure. III. Electronic Spectra and Electronic Structure of Polyatomic Molecules*. Van Nostrand Reinhold, New York (1966).

[55] H. H. Voge. *J. Chem. Phys.*, **4**, 581 (1936).

[56] A. Veillard. *Theoret. Chim. Acta*, **18**, 21 (1970).

[57] V. Pophristic and L. Goodman. *Nature*, **411**, 565 (2001).

[58] R. B. Woodward and R. Hoffman. *The Conservation of Orbital Symmetry*. Academic Press, New York (1970).

[59] M. Randić. in *Valence Bond Theory and Chemical Structure*. Ed. by D. J. Klein and N. Trinajstić. Elsevier, Amsterdam (1990).

[60] J. M. Norbeck and G. A. Gallup. *J. Amer. Chem. Soc.*, **96**, 3386 (1974).

[61] D. L. Cooper, J. Gerratt, and M. Raimondi. *Nature*, **323**, 699 (1986).

[62] S. Shaik, A. Shurki, D. Danovich, and P. C. Hiberty. *J. Mol. Struct. (THEOCHEM)*, **398**, 155 (1997).

[63] S. Shaik, A. Shurki, D. Danovich, and P. C. Hiberty. *Chem. Rev.* **101**, 1501 (2001).

[64] R. S. Mulliken and R. G. Parr. *J. Chem. Phys.*, **19**, 1271 (1951).

[65] H.-D. Beckhaus, R. Faust, A. J. Matzger, D. L. Mohler, D. W. Rogers, C. Rüchart, A. K. Sawhney, S. P. Verevkin, K. P. C. Vollhardt, and S. Wolff. *J. Amer. Chem. Soc.*, **122**, 7819 (2000).

[66] J. R. Collins and G. A. Gallup. *J. Amer. Chem. Soc.*, **104**, 1530 (1982).

[67] S. Shaik and A. Shurki. *Angew. Chem. Int. Ed.*, **38**, 586 (1999).

[68] R. L. Vance and G. A. Gallup, *Chem. Phys. Lett.*, **81**, 98 (1981).

[69] S. S. Shaik, H. B. Schlegel, and S. Wolfe. *Theoretical Aspects of Physical Organic Chemistry. The $S_N2$ Mechanism*. Wiley and Sons Inc., New York (1992).

[70] T. H. Dunning. *J. Chem. Phys.*, **65**, 2823 (1970).

[71] W. J. Hehre, L. Radom, P. v.R. Schleyer, and J. A. Pople. *Ab Initio Molecular Orbital Theory*. Wiley and Sons, New York (1986).

# Index